COBIT Focus Area: Information & Technology Risk

Using COBIT 2019

COBIT®
An ISACA® Framework

ISACA®

About ISACA

For more than 50 years, ISACA® (*www.isaca.org*) has advanced the best talent, expertise and learning in technology. ISACA equips individuals with knowledge, credentials, education and community to progress their careers and transform their organizations, and enables enterprises to train and build quality teams that effectively drive IT audit, risk management and security priorities forward. ISACA is a global professional association and learning organization that leverages the expertise of more than 150,000 members who work in information security, governance, assurance, risk and privacy to drive innovation through technology. It has a presence in 188 countries, including more than 220 chapters worldwide. In 2020, ISACA launched One In Tech, a philanthropic foundation that supports IT education and career pathways for under-resourced, under-represented populations.

Disclaimer

ISACA has designed and created *COBIT Focus Area: Information & Technology Risk* (the "Work") primarily as an educational resource for enterprise governance of information and technology (EGIT), assurance, risk and security professionals. ISACA makes no claim that use of any of the Work will assure a successful outcome. The Work should not be considered inclusive of all proper information, procedures and tests or exclusive of other information, procedures and tests that are reasonably directed to obtaining the same results. In determining the propriety of any specific information, procedure or test, enterprise governance of information and technology (EGIT), assurance, risk and security professionals should apply their own professional judgment to the specific circumstances presented by the particular systems or information technology environment.

ISACA
1700 E. Golf Road, Suite 400
Schaumburg, IL 60173, USA
Phone: +1.847.660.5505
Fax: +1.847.253.1755
Contact us: https://support.isaca.org
Website: www.isaca.org

Participate in the ISACA Online Forums: https://engage.isaca.org/onlineforums

Twitter: http://twitter.com/ISACANews
LinkedIn: www.linkedin.com/company/isaca
Facebook: www.facebook.com/ISACAGlobal
Instagram: www.instagram.com/isacanews/

COBIT Focus Area: Information & Technology Risk
ISBN 978-1-60420-866-5

Acknowledgments

ISACA wishes to recognize:

Lead Developers

Mattias Goorden, PwC, Belgium
Stefanie Grijp, Belgium
Dirk Steuperaert, CISA, CRISC, CGEIT, IT in Balance, Belgium

Expert Reviewers

Osman Azab, CISA, CRISC, CISM, CGEIT, Arab African International Bank, Egypt
Leela Ravi Shankar Dhulipalla, CGEIT, CDPSE, COBIT Assessor, TOGAF, PMP, NCSP Practitioner, IAITAM–CHAMP, CSAM, CITAM, CDMP Associate, AWS Cloud Practitioner, Accredited Trainer–COBIT, CGEIT & IAITAM, First Abu Dhabi Bank, UAE
J. Winston Hayden, CISA, CRISC, CISM, CGEIT, CDPSE, South Africa
Mike Hughes, CISA, CRISC, CISM, CGEIT, CDPSE, MIoD, Prism RA, United Kingdom
Hussam Khattab, CISA, CGEIT, CDPSE, PMP, COBIT2019 Design & Implementation, ITIL, ISACA Amman Chapter, Jordan
Linda Kostic, CISA, CISSP, CPA, Doctor of IT-Cybersecurity & Information Assurance, PRMIA Complete Course in Risk Management, George Washington University, Citi, USA
Lilia Liu, CRISC, CDPSE, CFE (from ACFE) and COBIT 5 Accredited by APMG, COSO ERM, Moore Panama, Panama
Salah Eddine Mahrach, Morocco
Larry Marks, CISA, CRISC, CISM, CGEIT, CDPSE, CISSP, ITIL, PMP, BDO USA LLP, USA
Ilker Tutu, CISA, CRISC, CISM, CGEIT, CCSP, CISSP, CIA, PCI ISA, PCIP, PayPal Europe, Luxembourg
Lisa Young, CISA, CISM, CISSP, Axio, USA
Dušan Žikić, CISA, CRISC, CISM, CSX-F, CSX-P, Cybersecurity Audit Certificate, COBIT 5 Foundation, COBIT 2019 Foundation, COBIT 2019 Design and Implementation, CISSP, CFE, CAPM, ITIL v4 Foundation, NIS Gazprom Neft, Serbia
Tichaona Zororo, CISA, CRISC, CISM, CGEIT, CDPSE, CIA, CRMA, COBIT Certified Assessor, EGIT | Enterprise Governance of IT, South Africa

Board of Directors

Tracey Dedrick, Chair, Former Chief Risk Officer, Hudson City Bancorp, USA
Rolf von Roessing, Vice-Chair, CISA, CISM, CGEIT, CDPSE, CISSP, FBCI, Partner, FORFA Consulting AG, Switzerland
Gabriela Hernandez-Cardoso, Independent Board Member, Mexico
Pam Nigro, CISA, CRISC, CGEIT, CRMA, Vice President–Information Technology, Security Officer, Home Access Health, USA
Maureen O'Connell, Board Chair, Acacia Research (NASDAQ), Former Chief Financial Officer and Chief Administration Officer, Scholastic, Inc., USA
David Samuelson, Chief Executive Officer, ISACA, USA
Gerrard Schmid, President and Chief Executive Officer, Diebold Nixdorf, USA
Gregory Touhill, CISM, CISSP, President, AppGate Federal Group, USA
Asaf Weisberg, CISA, CRISC, CISM, CGEIT, Chief Executive Officer, introSight Ltd., Israel
Anna Yip, Chief Executive Officer, SmarTone Telecommunications Limited, Hong Kong
Brennan P. Baybeck, CISA, CRISC, CISM, CISSP, ISACA Board Chair, 2019-2020, Vice President and Chief Information Security Officer for Customer Services, Oracle Corporation, USA
Rob Clyde, CISM, ISACA Board Chair, 2018-2019, Independent Director, Titus, and Executive Chair, White Cloud Security, USA
Chris K. Dimitriadis, Ph.D., CISA, CRISC, CISM, ISACA Board Chair, 2015-2017, Group Chief Executive Officer, INTRALOT, Greece

Page intentionally left blank

TABLE OF CONTENTS

LIST OF FIGURES

Chapter 1. Introduction

Chapter 2. Structure of This Publication and Intended Audience

Chapter 3. Structure of COBIT Governance and Management Objectives

Appendix A. I&T Risk Scenarios

Appendix C. Template for Risk Register Entry

Appendix D. IT Risk Reporting Examples

Appendix E. Sample Risk Maps

Chapter 1
Introduction

1.1 COBIT as an Information and Technology (I&T) Governance Framework

Over the years, best practice frameworks have been developed and promoted to assist in the process of understanding, designing and implementing enterprise governance of IT (EGIT). COBIT® 2019 builds on and integrates more than 25 years of development in this field, not only incorporating new insights from science but also operationalizing these insights as practices.

From its foundation in the IT audit community, COBIT® has developed into a broader and more comprehensive information and technology (I&T) governance and management framework and continues to establish itself as a generally accepted framework for I&T governance. Throughout this framework, IT is used to refer to the enterprise function with the primary responsibility for technology. I&T, as used in this framework, refers to all the information the enterprise generates, processes and uses to achieve its goals, as well as the supporting technologies.

1.1.1 What Is COBIT and What Is It Not?

Before describing the COBIT framework, it is important to explain what COBIT is and is not.

COBIT is a framework for the governance and management of information and technology, aimed at the whole enterprise. Enterprise I&T includes all the technology and information processing the enterprise puts in place to achieve its goals, regardless of where this happens in the enterprise. In other words, enterprise I&T is not limited to the IT department of an organization but certainly includes it.

The COBIT framework makes a clear distinction between governance and management. These two disciplines encompass different activities, require different organizational structures and serve different purposes.

- Governance ensures that:
    - Stakeholder needs, conditions and options are evaluated to determine balanced, agreed-on enterprise objectives.
    - Direction is set through prioritization and decision making.
    - Performance and compliance are monitored against agreed-on direction and objectives.

In most enterprises, governance is the responsibility of the board of directors, under the leadership of the chairperson. Specific governance responsibilities may be delegated to special organizational structures at an appropriate level, particularly in larger, complex enterprises.

- Management plans, builds, runs and monitors activities, in alignment with the direction set by the governance body, to achieve enterprise objectives.

In most enterprises, management is the responsibility of the executive management, under the leadership of the chief executive officer (CEO).

COBIT defines the components to build and sustain a governance system: processes, organizational structures, principles, policies and procedures, information flows, culture and behaviors, skills, and infrastructure.

COBIT defines the design factors that should be considered by the enterprise to build a best-fit governance system.

COBIT addresses governance issues by grouping relevant governance components into governance and management objectives that can be managed to the required capability levels.

Several misconceptions about COBIT should be dispelled:

- COBIT is not a full description of the whole IT environment of an enterprise.
- COBIT is not a framework to organize business processes.
- COBIT is not an (IT-)technical framework to manage all technology.

COBIT does not make or prescribe any IT-related decisions. It will not decide what the best IT strategy is, what the best architecture is, or how much IT can or should cost. Rather, COBIT defines all the components that describe which decisions should be taken, and how and by whom they should be taken.

1.2 COBIT Overview

The COBIT 2019 product family is open-ended and designed for customization. The following publications are currently available:

- ***COBIT® 2019 Framework: Introduction and Methodology***[1] introduces the key concepts of COBIT 2019.
- ***COBIT® 2019 Framework: Governance and Management Objectives***[2] comprehensively describes the 40 core governance and management objectives, the processes contained therein, and other related components. This guide also references other standards and frameworks.
- ***COBIT® 2019 Design Guide: Designing an Information and Technology Governance Solution***[3] explores design factors that can influence governance and includes a workflow for planning a tailored governance system for the enterprise.
- ***COBIT® 2019 Implementation Guide: Implementing and Optimizing an Information and Technology Governance Solution***[4] represents an evolution of the *COBIT® 5 Implementation Guide* and develops a road map for continuous governance improvement. It may be used in combination with the COBIT 2019 Design Guide.

Figure 1.1 shows the high-level overview of COBIT 2019 and illustrates how different publications within the set cover different aspects.

[1] ISACA, *COBIT® 2019 Framework: Introduction and Methodology*, USA, 2018, *www.isaca.org/bookstore/bookstore-cobit_19-print/cb19fim*

[2] ISACA, *COBIT® 2019 Framework: Governance and Management Objectives*, USA, 2018, *www.isaca.org/bookstore/bookstore-cobit_19-print/cb19fgm*

[3] ISACA, *COBIT® 2019 Design Guide: Designing an Information and Technology Governance Solution*, USA, 2018, *www.isaca.org/bookstore/bookstore-cobit_19-digital/wcb19dgd*

[4] ISACA, *COBIT® 2019 Implementation Guide: Implementing and Optimizing an Information and Technology Governance Solution*, USA, 2018, *www.isaca.org/bookstore/bookstore-cobit_19-print/cb19igio*

Figure 1.1—COBIT Overview

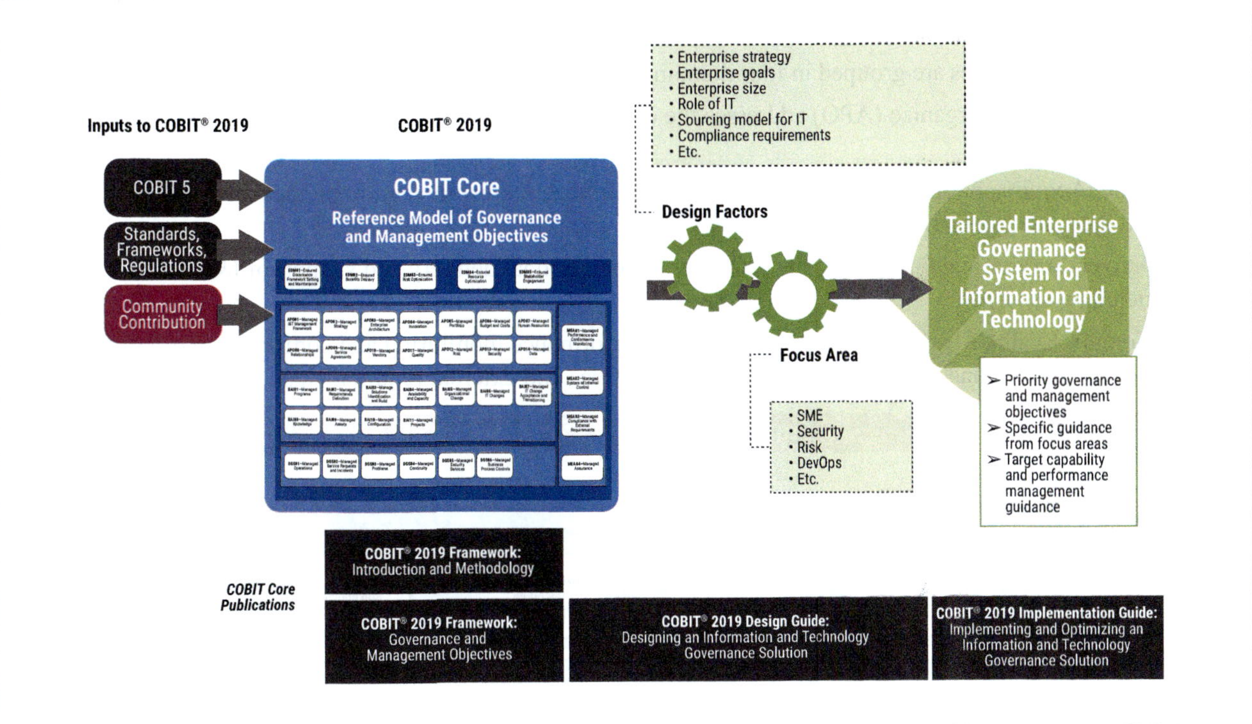

Source: ISACA, *COBIT® 2019 Framework: Introduction and Methodology*, USA, 2018, figure 4.1, *www.isaca.org/bookstore/bookstore-cobit_19-print/cb19fim*

The following sections explain the key concepts and terms used in COBIT 2019.

Focus area content (see **figure 1.1**) contains more detailed guidance on specific themes. This focus area publication provides a tailored version of guidance—specific to I&T risk—based on (and extending from) the core publication, *COBIT® 2019 Framework: Governance and Management Objectives.*

1.3 Terminology and Key Concepts of the COBIT Framework

1.3.1 Governance and Management Objectives

For information and technology to contribute to enterprise goals, several governance and management objectives should be achieved. Basic concepts relating to governance and management objectives are:

- A governance or management objective always relates to one process (with an identical or similar name) and a series of related components of other types to help achieve the objective.
- A governance objective relates to a governance process (depicted on the dark blue background in **figure 1.2**), while a management objective relates to management processes (depicted on the lighter blue background in **figure 1.2**). Boards and executive management are typically accountable for governance processes, while management processes are the domain of senior and middle management.

The governance and management objectives in COBIT are grouped into five domains. The domains have names with verbs that express the key purpose and areas of activity of the objectives contained in them:

- Governance objectives are grouped in the Evaluate, Direct and Monitor (EDM) domain. In this domain, the governing body evaluates strategic options, directs senior management on the chosen strategic options and monitors the achievement of the strategy.
- Management objectives are grouped in four domains:
  - Align, Plan and Organize (APO) addresses the overall organization, strategy and supporting activities for I&T.
  - Build, Acquire and Implement (BAI) treats the definition, acquisition and implementation of I&T solutions and their integration in business processes.
  - Deliver, Service and Support (DSS) addresses the operational delivery and support of I&T services, including security.
  - Monitor, Evaluate and Assess (MEA) addresses performance monitoring and conformance of I&T with internal performance targets, internal control objectives and external requirements.

Figure 1.2—COBIT Core Model

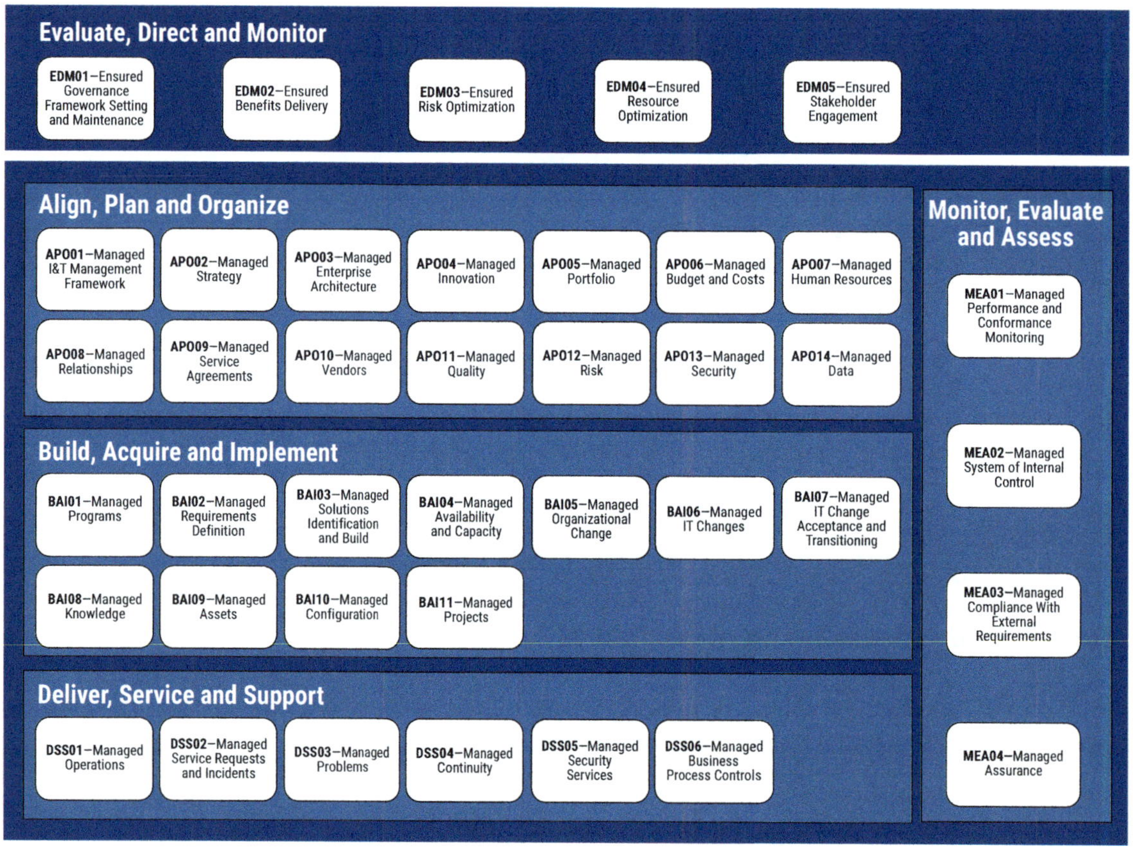

Source: ISACA, *COBIT® 2019 Framework: Introduction and Methodology*, USA, figure 4.2, *www.isaca.org/bookstore/bookstore-cobit_19-print/cb19fim*

1.3.2 Components of the Governance System

To satisfy governance and management objectives, each enterprise needs to establish, tailor and sustain a governance system built from several components:

- Components are factors that, individually and collectively, contribute to the good operations of the enterprise's governance system over I&T.
- Components interact with each other, resulting in a holistic governance system for I&T.
- Components can be of different types. The most familiar are processes. However, components of a governance system also include organizational structures; principles, policies and procedures; information flows and items; culture, ethics and behavior; people, skills and competencies; and services, infrastructure and applications (**figure 1.3**).
  - Processes describe an organized set of practices and activities to achieve certain objectives and produce a set of outputs that support achievement of overall IT-related goals.
  - Organizational structures are the key decision-making entities in an enterprise.
  - Principles, policies and procedures translate desired behavior into practical guidance for day-to-day management.
  - Information is pervasive throughout any organization and includes all information produced and used by the enterprise. COBIT focuses on the information required for the effective functioning of the governance system of the enterprise.
  - Culture, ethics and behavior of individuals and of the enterprise are often underestimated as factors in the success of governance and management activities.
  - People, skills and competencies are required for good decisions, execution of corrective action and successful completion of all activities.
  - Services, infrastructure and applications include the infrastructure, technology and applications that provide the enterprise with the governance system for I&T processing.

Figure 1.3—COBIT Components of a Governance System

Source: ISACA, *COBIT 2019 Framework: Introduction and Methodology*, USA, 2018, figure 4.3, *www.isaca.org/bookstore/bookstore-cobit_19-print/cb19fim*

Components of all types can be generic or can be variants of generic components:

- Generic components are described in the COBIT core model (**figure 1.2**) and apply in principle to any situation. However, they are generic in nature and generally need customization before being practically implemented.
- Variants are based on generic components but are tailored for a specific purpose or context within a focus area (e.g., for information security, DevOps, a particular regulation).

1.3.3 Focus Areas

A focus area describes a certain governance topic, domain or issue that can be addressed by a collection of governance and management objectives and their components. Examples of potential focus areas include small and medium enterprises, digital transformation, cloud computing, privacy, and DevOps.[5]

Focus areas may contain a combination of generic governance components and variants.

This publication shows how COBIT 2019 can be tailored as an I&T risk framework and system and indicates where I&T risk is relevant in the COBIT governance and management framework. It looks at the COBIT concepts from an I&T risk perspective and illustrates this perspective using COBIT to design and implement, govern and manage I&T risk capabilities in the enterprise. This publication is not intended to be used to perform I&T risk assessments. For guidance on how to perform a risk assessment, see ISACA's *Risk IT Practitioner Guide*.[6]

The number of focus areas is virtually unlimited. That is what makes COBIT open-ended. New focus areas can be added as required or as subject matter experts and practitioners contribute to the open-ended COBIT model.

1.4 I&T Risk Focus Area Overview

This publication provides specific guidance related to the I&T risk focus area (IRFA). It is based on the COBIT core guidance for governance and management objectives, and it has been tailored to address I&T risk practices.

In formulating a business or operational strategy, the board of directors and senior management often decide explicitly to accept some level of risk to achieve enterprise objectives. In COBIT, this practice is known as risk appetite, that is, the amount of risk, on a broad level, that an entity is willing to accept in pursuit of its mission (or vision) and the achievement of business objectives. Risk governance and management primarily focus on leveraging resources and activities to help reduce the business impact from a realized risk or the likelihood (or probability) of a risk materializing that exceeds acceptable levels. The framework broadly facilitates management of the entire spectrum of I&T-related risk; however, as relevant subcategories, information security[7] and cyberrisk examples may be used to show the interrelated nature of systems and processes.

1.4.1 Risk Fundamentals

Risk is the combination of the likelihood of an event and its impact. Risk is most often associated with uncertainties and deviations from expected results that can be harmful to an enterprise and threaten its ability to meet its business

[5] See, for example, ISACA, *COBIT Focus Area: Information Security*, USA 2020, *www.isaca.org/bookstore/bookstore-cobit_19-print/cb19is*

[6] ISACA, *Risk IT Practitioner Guide, 2nd Edition*, USA, 2020, *www.isaca.org/bookstore/bookstore-risk-print/ritpg2*

[7] Some people use the terms information security and cybersecurity interchangeably. Cybersecurity is actually a component of information security. The goal of information security is to keep data safe from unauthorized access, modification and theft during processing, storage and transmission. Information security deals with all formats of information—paper documents, digital assets, intellectual property, and verbal and visual communications. Cybersecurity is the protection of information assets by addressing threats to information processed, stored and transported by internetworked information systems. This publication uses the broader term information security, but readers should note that this term encompasses cybersecurity by definition.

objectives. This publication defines I&T risk as business risk, specifically, the business risk associated with the use, ownership, operation, involvement, influence and adoption of information and technology.

I&T risk always exists, whether it is detected or recognized by an enterprise. Every business opportunity could involve some risk. As the enterprise embarks upon conducting business in pursuit of its objectives and goals, it will engage in activities that involve many resources, including technology. With these engagements, unknown factors always exist. Future events cannot be predicted with complete certainty. Business plans and strategies developed for the future include assumptions, estimates and calculations. There is always the potential that future occurrences will be different from the expected and the possibility of unfavorable outcomes exists.

1.4.2 Drivers

The main drivers for risk management in its different forms include the need to improve business outcomes, decision making and overall strategy by providing:

- Stakeholders with substantiated data on the current state of risk throughout the enterprise
- Guidance on how to manage the risk to levels within the risk appetite and risk tolerance of the enterprise
- Guidance on how to set up the appropriate risk culture for the enterprise
- Wherever possible, quantitative risk assessments that enable stakeholders to consider the cost of mitigation and the required resources against the loss exposure

To that purpose, this publication provides:

- Guidance on how to use the COBIT core model to establish the I&T risk governance and management functions for the enterprise
- Guidance and a structured approach on how to use the COBIT principles to govern and manage I&T risk

1.4.3 Benefits

This focus area can help an enterprise increase its ability to govern and manage I&T risk and provides other benefits, including:

- More accurate identification of risk and measurement of success in addressing that risk
- Better understanding of risk impact on the enterprise
- End-to-end guidance on how to manage risk, including an extensive set of measures
- Knowledge of how to capitalize on investments related to I&T risk management practices
- Understanding of how effective I&T risk management optimizes value, with business process effectiveness and efficiency, improved quality, and reduced waste and costs
- Opportunities to integrate I&T risk management with enterprise risk and compliance requirements
- Improved communication and understanding among all internal and external stakeholders due to the common and sustainable globally accepted framework and language for assessing and responding to risk
- Promotion of risk responsibility and acceptance across the enterprise
- A complete risk profile, identifying the full enterprise risk exposure and enabling better utilization of enterprise resources
- Improved risk awareness throughout the enterprise

Page intentionally left blank

Chapter 2

Structure of This Publication and Intended Audience

2.1 Structure of This Publication

This publication provides a comprehensive description of the 23 of 40 governance and management objectives relevant to I&T risk, as defined in the COBIT core model, shown in figure 1.2.

The balance of this publication includes the following sections and appendices:

- Chapter 3 explains the structure that is used to detail the guidance for the 23 governance and management objectives, across all applicable governance components.
- Chapter 4 provides comprehensive, detailed I&T information risk-specific guidance for the 23 relevant governance and management objectives, focusing on two key components: processes (core model Component A) and information flows and items (core model Component C).

The 23 relevant governance and management components for I&T risk are:

Objective ID	I&T Risk-Relevant COBIT Objectives
EDM01	Ensured Governance Framework Setting and Maintenance
EDM02	Ensured Benefits Delivery
EDM03	Ensured Risk Optimization
EDM05	Ensured Stakeholder Engagement
APO01	Managed I&T Management Framework
APO02	Managed Strategy
APO05	Managed Portfolio
APO06	Managed Budget and Costs
APO07	Managed Human Resources
APO08	Managed Relationships
APO10	Managed Vendors
APO11	Managed Quality
APO14	Managed Data
BAI03	Managed Solutions Identification and Build
BAI05	Managed Organizational Change
BAI08	Managed Knowledge
DSS01	Managed Operations
DSS04	Managed Continuity
DSS06	Managed Business Process Controls
MEA01	Managed Performance and Conformance Monitoring
MEA02	Managed System of Internal Control
MEA03	Managed Compliance With External Requirements
MEA04	Managed Assurance

2.2 Intended Audience

The intended audience for *COBIT Focus Area: Information & Technology Risk* is extensive, as are the reasons for adopting and using the framework and the benefits that each enterprise role and function can find in this publication. The roles and functions that are listed in **figure 2.1** are considered key stakeholders in the risk management process. These stakeholders do not necessarily refer to individuals, but to roles and functions within the enterprise or its environment.

Figure 2.1—*COBIT I&T Risk Focus Area* Target Audience and Benefits	
Role/Function	**Benefit of/Reason for Adopting and Adapting *COBIT Focus Area: I&T Risk***
Board and executive management	Better understanding of their responsibilities and roles regarding I&T risk management, implications of I&T risk for enterprise strategic objectives and optimizing IT for successful strategy execution
Risk function and corporate risk managers for enterprise risk management (ERM)	Assistance with managing I&T risk, according to generally accepted ERM principles, and incorporating I&T risk into enterprise risk
Operational risk managers	Linkage of their framework to *COBIT Focus Area: I&T Risk* and identification of operational losses or development of key risk indicators (KRIs)
I&T management	Better understanding of how to identify and manage I&T risk and how to communicate I&T risk to business decision-makers
I&T service managers	Enhancement of their view of operational risk, which should fit into an overall I&T risk management framework
Business continuity	Alignment with ERM because assessment of risk is a key aspect of their responsibility
I&T security	Positioning security risk among other categories of I&T risk
Information security	Positioning I&T risk within the enterprise information risk management structure
Chief financial officer (CFO)	Gaining a better view of I&T risk and its financial implications for investment and portfolio management purposes
Enterprise governance officers	Assistance with their review and monitoring of governance responsibilities and other I&T governance roles
Business	Understanding and management of I&T risk—one of many business risk items, all of which should be managed consistently
Internal auditors	Improved analysis of risk in support of audit plans and reports
Compliance	Support with the role as key advisors to the risk function regarding compliance requirements and their potential impact on the enterprise
General counsel	Support with the role as key advisor for the risk function on regulation-related risk and potential impact or legal implications
Regulators	Support of their assessment of the regulated enterprise I&T risk management approach and the impact of risk on regulatory requirements
External auditors	Additional guidance on exposure levels when establishing an opinion on the quality of internal control
Insurers	Support with establishing adequate I&T insurance coverage and seeking agreement on exposure levels
Rating agencies	In collaboration with insurers, a reference to assess and rate objectively how an enterprise is managing I&T risk
I&T contractors and subcontractors	Better alignment of utility and warranty of I&T services and better understanding of responsibilities arising from risk assessment

Chapter 3
Structure of COBIT Governance and Management Objectives

3.1 Introduction

This chapter describes the generic structure used to detail each of the in-scope COBIT I&T governance and management objectives. For each I&T risk governance and management objective, chapter 4 of this publication provides information related to the following two governance components applicable to that governance or management objective:

- Processes (section 3.3)
- Information flows and items (section 3.5)

The structure for this information is detailed in sections 3.2, 3.3 and 3.5. For each governance and management objective, chapter 4 elaborates additional detail specific to the I&T risk focus area for the process and information flows and items components (Components A and C, respectively, in the COBIT core model).

Additional information for the other five components is detailed in this chapter. These components include:

- Organizational structures (section 3.4)
- People, skills and competencies (section 3.6)
- Principles, policies and procedures (section 3.7)
- Culture, ethics and behavior (section 3.8)
- Services, infrastructure and applications (section 3.9)

3.2 Governance and Management Objectives

COBIT 2019 includes 40 governance and management objectives, organized into five domains (**figure 1.2**).

- There is one governance domain:
    - Evaluate, Direct and Monitor (EDM)
- There are four management domains:
    - Align, Plan and Organize (APO)
    - Build, Acquire and Implement (BAI)
    - Deliver, Service and Support (DSS)
    - Monitor, Evaluate and Assess (MEA)

The high-level information detailed for each objective (**figure 3.1**) includes:

- Domain name
- Focus area (in this publication, I&T risk)
- Governance or management objective name
- Description
- Purpose statement
- Focus area relevance statement

Focus area relevance statements describe how COBIT objectives relate to I&T risk.

Figure 3.1—Display of Governance and Management Objectives	
Domain: <NAME> Governance/Management Objective: <NAME>	Focus Area: I&T Risk
Description	
<TEXT>	
Purpose	
<TEXT>	
I&T Risk Focus Area Relevance	
<TEXT>	

In *COBIT® 2019 Framework: Governance and Management Objectives*, the governance objective EDM03 *Ensured Risk Optimization* and management objective APO12 *Managed Risk* already describe practices for risk governance and risk management in general.

Although 23 COBIT objectives are deemed most relevant to the I&T risk concerns of the entire enterprise, it is not necessarily the case that every governance and management objective must be adopted. Therefore, risk-relevant metrics and activities are provided for 23 of 40 overall COBIT objectives. Metrics and activities outlined in chapter 4 should be consulted (in addition to the COBIT 2019 core model guidance) when practitioners apply COBIT in an I&T risk-related context.

3.3 Component: Process

Each governance and management objective includes a process component with several governance or management process practices (**figure 3.2**). Each practice has one or more activities. A number of example metrics accompanies each process practice, to measure the achievement of the process practice and its contribution to the achievement of the overall objective. metrics in this publication relate specifically to I&T risk, and they should be considered in addition to the core model metrics in *COBIT® 2019 Framework: Governance and Management Objectives.*

I&T risk-specific activities are included in each practice. Several activities specific to I&T risk supplement governance objective EDM03 *Ensured Risk Optimization* and management objective APO12 *Managed Risk*. Users should also consult the core model, which provides basic guidance to define, operate and monitor a system for general security.

Figure 3.2—Display of Process Component		
A. Component: Process		
Governance/Management Practice	**Example I&T Risk-specific Metrics**	
<REF> <NAME> <DESCRIPTION>	<METRIC>	
I&T Risk-specific Activities (in addition to COBIT 2019 core-model activities)		**Capability Level**
1. <TEXT>		<NR>
2. <TEXT>		<NR>
n. <TEXT>		<NR>
Related Guidance (Standards, Frameworks, Compliance Requirements)	**Detailed Reference**	
<STANDARD NAME>	<TEXT>	

A capability level is assigned to all process activities, enabling clear definition of processes at different capability levels. A process reaches a certain capability level when all activities at that capability level are performed successfully. COBIT supports a CMMI-based process capability scheme, ranging from 0 to 5. The capability level is a measure of how well a process is implemented and performing. **Figure 3.3** depicts the model, the increasing capability levels and the general characteristics of each.

Figure 3.3—Maturity Levels for Focus Areas

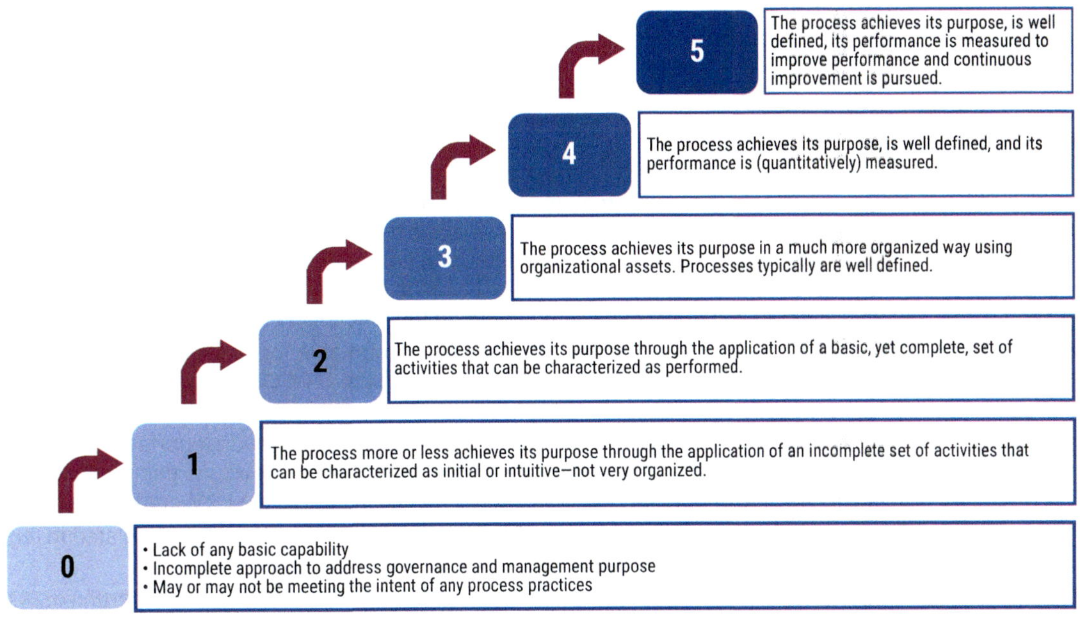

Source: ISACA, *COBIT® 2019 Framework: Introduction and Methodology*, USA, 2018, figure 6.3, *www.isaca.org/bookstore/bookstore-cobit_19-print/cb19fim*

References to other standards and guidance are offered for relevant practices (**figure 3.2**). Related guidance refers to all standards, frameworks, compliance requirements and guidance that are relevant for the process at hand. Detailed references cite specific chapters or sections within related guidance. A complete list of sources for related guidance is included in Appendix C of *COBIT® 2019 Framework: Governance and Management Objectives.*

3.4 Component: Organizational Structures

3.4.1 Introduction

The organizational structures governance component is not reproduced in chapter 4 of this publication similar to its presentation in chapter 4 of the core model (*COBIT® 2019 Framework: Governance and Management Objectives*). A discussion, however, of organizational structures and their applicability to I&T risk follows.

The following roles and organizational structures—defined in the COBIT 2019 core model—are relevant for *COBIT Focus Area: Information & Technology Risk.*

Not every organization will have (or will need) all possible roles—and not all roles are essential for the adoption of the I&T risk focus area. Any risk-related functions within an enterprise may include roles described in **figure 3.4**.[8]

[8] Roles and descriptions are excerpted from ISACA, *COBIT® 2019 Framework: Governance and Management Objectives*, USA, 2018, Appendix B, *www.isaca.org/bookstore/bookstore-cobit_19-print/cb19fgm*. Using a standard responsible, accountable, consulted, informed (RACI) schema, *COBIT® 2019 Framework: Governance and Management Objectives* assigns roles to one of two categories (or levels) of participation: accountable (A), responsible (R). In addition to Appendix B, see also *COBIT® 2019 Framework: Governance and Management Objectives*, section 3.5 and chapter 4 *passim*.

3.4.2 Key Roles and Structures for the I&T Risk Function

Governance objective EDM03 *Ensured Risk Optimization* and management objective APO12 *Managed Risk* contemplate the use and optimization of several key I&T risk decision-making roles and organizational entities in an enterprise (note that some of these roles would be considered consulted or informed, as opposed to responsible or accountable).

Figure 3.4—COBIT Key Roles and Organizational Structures for I&T Risk Functions

Role/Structure	Description
Enterprise risk management (ERM) committee	The group of executives of the enterprise who are accountable for the enterprise-level collaboration and consensus required to support ERM activities and decisions; they are considered to be the second line of defense against risk manifestation. An IT risk council may be established to consider IT risk in more detail and advise the ERM committee. Committee members are usually drawn from the board and the CEO chairs the committee. The organization's internal audit manager may be invited to attend the meeting as a guest.
Chief risk officer (CRO)	The most senior official of the enterprise who is accountable for all aspects of risk management across the enterprise, including taking direction from the ERM Committee. An I&T risk officer function may be established to oversee risk.
Audit department	The function in the enterprise responsible for providing internal audit reports on the risk associated with gaps in controls identified while performing reviews. As this is considered the third and last line of defense, a representative can be invited to the ERM committee.
Compliance department	The function in the enterprise responsible for insight into the enterprise risk related to regulations, legal mandates, and internal policies and standards.
I&T risk officers/ managers	The group of risk professionals responsible for identifying, assessing and reporting of I&T risk. They provide oversight, challenge, assurance and support to I&T risk owners.

Figures 3.5 through **3.17** describe in more detail the mandate, operating principles, responsibilities, authority level, and delegation and escalation rights associated with roles and entities appropriate for enterprises of a certain size and complexity that want to manage I&T risk:

- Enterprise risk management (ERM) committee
- Chief risk officer (CRO)
- Audit department
- Compliance department
- I&T risk officers/managers

The detailed descriptions of these organizational structures are meant to be used as generic guidance or examples and therefore do not take into account enterprise-specific situations. For example, in the case of a parent company with multiple subsidiaries, an ERM committee may or may not exist in every subsidiary. The span of control of these subsidiary organizations may be limited within this specific organization, not covering all group subsidiaries.

The organization structures described are examples; they are not prescriptive. As a matter of fact, not all enterprises—depending on their nature, size or any other contextual circumstance—may have implemented all of these structures.

Enterprise Risk Management Committee Composition

Figure 3.5 describes the composition of the ERM committee by member roles and their function within the context of this structure.

Figure 3.5—Composition of the Enterprise Risk Management Committee	
Role	**Description**
Chief risk officer (CRO)	The most senior official of the enterprise who is accountable for all aspects of risk management across the enterprise. An I&T risk officer function may be established to oversee risk. In some organizations, the CEO is charged with chairing the committee, per delegation by the board, to oversee the day-to-day risk in the organization, when there is no specific CRO role.
Chief financial officer (CFO)	The most senior official of the enterprise who is accountable for all aspects of financial management, including financial risk and controls and reliable and accurate accounts
Chief information officer (CIO)	The most senior official of the enterprise who is responsible for aligning IT and business strategies and accountable for planning, resourcing and managing the delivery of IT services and solutions to support enterprise objectives
Chief operating officer (COO)	The most senior official of the enterprise who is accountable for the operations of the enterprise
Compliance department	The function in the enterprise responsible for guidance on legal, regulatory and contractual compliance
Audit department	The function in the enterprise responsible for conducting internal audit reviews and coordinating external audit
Privacy officer	An individual who is responsible for monitoring the risk and business impacts of privacy laws and for guiding and coordinating the implementation of policies and activities that will ensure that the privacy directives are met
Chief information security officer (CISO)	The most senior officer of the enterprise who is accountable for the security of enterprise information in all forms
HR department	The most senior official of an enterprise who is accountable for planning and policies with respect to all human resources in that enterprise
Business owner (as appropriate for agenda under review)	A senior management individual accountable for the operation of a specific business unit or subsidiary. This includes key business line owners and heads of departments such as sales, marketing, human resources, manufacturing, etc.

Figure 3.6 describes the boundaries of the ERM committee's activities and the authority level that is attributed to it.

Figure 3.6—ERM Committee Mandate, Operating Principles, Span of Control and Authority Level	
	Activities
Mandate	Assists the board and the audit committee in supervising the enterprise risk management (ERM) activities and advises the board with respect to the ERM framework
Operating principles	• The ERM committee will meet on a regular basis (e.g., quarterly). More frequent meetings may be scheduled during specific initiatives or when issues need to be dealt with on a very urgent basis. • The committee reports to the board on a regular basis (e.g., quarterly) or as needed and makes recommendations on the activities covered by its mandate. • The committee approves minutes, including the agenda, decisions made, attendance, action items and status reports. Approval may be given at a time such as at the beginning of the next meeting and the approved documents are retained in accordance with the enterprise retention policy. • Housekeeping rules: A chairperson will be named and the meeting will be run in accordance with a set of guiding principles such as Robert's Rules of Order.
Span of control	The ERM committee services the entire legal entity for which the board is responsible.

Figure 3.6—ERM Committee Mandate, Operating Principles, Span of Control and Authority Level *(cont.)*	
	Activities
Authority level/decision rights	The ERM committee is responsible to: • Direct the enterprise's risk management system, framework and methods (e.g., risk governance structure, assessment method, risk appetite) • Approve the ERM policy (under a delegation from the board) • Review the exposure • Oversee risk activities to confirm adherence to the board's defined risk appetite • Direct risk prioritization and strategy • Review the status of risk response and coordinate with business owners to allocate resources
Delegation rights	The ERM committee has the right to delegate risk ownership to risk owners/subcommittees or business owners, depending on the enterprise's size and complexity.
Escalation path	Since I&T risk management is considered the second line of defense, all key issues and findings impacting the board's direction need to be escalated to the CEO and board.

Responsible and Accountable (RA) Chart

Figure 3.7 lists process practices where the organizational structure carries responsibilities (R) and accountabilities (A). The level of approval within an enterprise is dependent on each enterprise's risk appetite or geographic regulatory requirements. **Figure 3.7** is a representative example.

Figure 3.7—RA Chart for Enterprise Risk Management Committee	
Process Practice	**Level of Involvement (R/A)**
Steer the risk management system, framework and methods (e.g., risk governance structure, assessment method and risk appetite)	A
Review and approve the enterprise risk management policy.	R
Review the exposure and tolerance, aligned with the enterprise's risk appetite.	A
Institutionalize the enterprise's risk appetite.	A
Steer and approve risk prioritization and strategy.	A
Review the status of risk closure and coordinate with business owners the appropriate allocation of resources.	A
Report the exposure to the board.	A

Inputs/Outputs

The following table lists the inputs the structure needs to fulfill the committee's mandate within the defined span of control. Inputs can consist of particular items of information, policies or procedures, or decisions from other structures.

Figure 3.8—Inputs for Enterprise Risk Management Committee		
Input	**Type**	**From**
Key risk indicators (KRIs), key performance indicators (KPIs) and key goal indicators (KGIs)	Information	Performance reporting
Incident reports	Information	Incident process
Business strategy (e.g., emerging technologies)	Information	Strategy process
Policies	Decision	Enterprise governance

Figure 3.8—Inputs for Enterprise Risk Management Committee *(cont.)*		
Input	**Type**	**From**
Audit reports or other reviews	Information	Audit reviews (internal and external) Security assessments and tests, Binding Corporate Roles (BCRs) tests
Risk report (current and mitigation status)	Decision	Risk management
Risk register	Information	Risk management
Regulations	Information	Legal and compliance
Threat intelligence	Information	Threat intelligence providers (internal and external)

Figure 3.9 lists the outputs the committee will generate by executing the mandate within the defined span of control. Outputs can be decisions, particular pieces of information or policies/procedures.

Figure 3.9—Outputs for Enterprise Risk Management Committee		
Output	**Type**	**To**
Risk tolerance	Information	Risk management
Risk appetite	Information	Risk management
Meeting minutes	Decision	Risk management Externals, as appropriate
Risk management strategy	Decision	Board Business owners
Risk mitigation actions	Decision	Board Business owners IT process owners
Policy (change control)	Decision	Communications

Chief Risk Officer

Figure 3.10 describes the boundaries of the CRO's activities and the authority level that is attributed to this role.

Figure 3.10—CRO Mandate, Operating Principles, Span of Control and Authority Level	
	Activities
Mandate	Overall responsibility for the development and implementation of the enterprise risk management (ERM) program
Operating principles	Ensures a holistic approach to risk management. The CRO sets the parameters of the risk framework, maintains the risk profile and reports on significant risk issues to the CEO and the ERM committee on a regular basis.
Span of control	The CRO's span of control is the entire enterprise risk environment.
Authority level/decision rights	Directs and manages the ERM program
Delegation rights	The CRO assigns risk ownership and responsibilities.
Escalation path	Escalates to the CEO and ERM committee

RA Chart

Figure 3.11 lists process practices for the CRO. The level of approval within an enterprise is dependent on each enterprise's risk appetite or geographic regulatory requirements. **Figure 3.11** is a representative example.

Figure 3.11—RA Chart for Chief Risk Officer

Process Practice	Level of Involvement (R/A)
Research, define and document the enterprise's risk management requirements.	R
Define, manage and continuously improve the enterprise risk management (ERM) system, method, framework and tools to help ensure consistent and effective risk acceptance decisions.	R
Develop the ERM strategy and implement the plan.	R
Develop risk management policy, procedures and other documents/templates.	A
Implement and manage the risk management process (EDM03) and the supporting tools and knowledge repositories.	R
Collect and communicate threats and vulnerability intelligence.	R
Provide subject matter expertise to identify current risk and projected risk associated with changes in the business strategy.	A
Provide a system to measure and report the enterprise loss exposure while aligning to the business strategy and managing risk response effectively.	R/A
Establish a communications strategy to promote collaboration, understanding of risk ownership and the desired behaviors with stakeholders across the enterprise.	R/A
Define a risk management action portfolio and manage opportunities to reduce risk to an acceptable level.	A

Audit Department

Figure 3.12 describes the boundaries of the audit department's activities and the authority level that is attributed to this structure.

Figure 3.12—Audit Department Mandate, Operating Principles, Span of Control and Authority Level

	Activities
Mandate	Responsible for testing, examination and reporting of control conditions within the enterprise. The mandate will be determined in the audit charter; the department will define and execute an audit plan usually based on business risk considerations.
Operating principles	Independence from operational responsibilities: Auditors must ensure their independence is unquestionable when performing their audit duties but, within this limitation, may give advice on, for example, business-related risk or business impact analysis (BIA).
Span of control	The span of control is enterprisewide. The audit department services the entire legal entity for which the board is responsible.
Authority level/decision rights	• Selecting audit processes and tools • Establishing risk ratings for audit findings • Direct reporting to the board
Delegation rights	The audit department may delegate some testing activities to other parts of the organization or external subject matter experts if it can assure the quality and reliability of the tests.
Escalation path	Audit committee, the board and/or CEO

RA Chart

Figure 3.13 offers a representative example of process practices for the audit department as they relate to I&T risk. The level of approval within an enterprise is dependent on each enterprise's risk appetite or geographic regulatory requirements.

Figure 3.13—RA Chart for Audit	
Process Practice	**Level of Involvement (R/A)**
Develop and execute the enterprise audit plan in alignment with the enterprise strategy.	R/A
Plan and perform audits.	R/A
Communicate findings and validate resulting remediation plans.	R/A
Audit the findings' severity settings.	R/A
Report the enterprise's control state to the board.	R/A
Submit an audit plan for approval by the board.	R/A
Engage with stakeholders.	R/A

Compliance Department

Figure 3.14 describes the boundaries of the compliance department's activities and the authority level that is attributed to it.

Figure 3.14—Compliance Department Mandate, Operating Principles, Span of Control and Authority Level	
	Activities
Mandate	Monitor and report on the state of compliance within the enterprise related to relevant laws and regulations.
Operating principles	Ensure that compliance requirements are understood, communicated and adhered to, and areas of noncompliance are remediated.
Span of control	The span of control is enterprisewide.
Authority level/decision rights	• Set up a compliance program. • Recommend a course of action to appropriate decision-makers. • Assign ownership of compliance activities.
Delegation rights	N/A
Escalation path	General counsel, CEO, enterprise risk management committee and/or board.

<u>RA Chart</u>

Figure 3.15 offers a representative example of process practices for which the compliance department carries responsibilities. The level of approval within an enterprise is dependent on each enterprise's risk appetite or geographic regulatory requirements.

Figure 3.15—RA Chart for the Compliance Department	
Process Practice	**Level of Involvement (R/A)**
Monitor relevant laws and regulations.	R/A
Interpret and communicate requirements from relevant laws and regulations to affected stakeholders.	R/A
Facilitate compliance activities.	R/A
Collaborate with regulatory bodies.	R/A
Facilitate regulatory examinations.	R/A
Perform internal compliance audits.	R/A
Communicate compliance examination results to appropriate stakeholders.	R/A
Make recommendations for achieving and maintaining compliance.	R/A
Understand the impact of noncompliance on the organization.	R/A

I&T Risk Officer/Manager

Figure 3.16 describes the boundaries of the structure's roles' activities and the authority level that is attributed to them.

Figure 3.16—Risk Officer/Manager Mandate, Operating Principles, Span of Control and Authority Level

	Activities
Mandate	Overall responsibility for the development and implementation of the enterprise risk management program
Operating principles	Ensure a holistic approach to risk management. These roles set the parameters of the risk framework, maintain the risk profile and report on significant risk issues to the CEO and enterprise risk management (ERM) committee on a regular basis.
Span of control	The I&T risk officer's/manager's span of control is the entire enterprise risk environment.
Authority level/decision rights	Direct and manage the ERM program
Delegation rights	Consult with business stakeholders on risk ownership and related responsibilities
Escalation path	Escalates to the CRO

Figure 3.17 lists process practices for the I&T risk officer/manager.

Figure 3.17—RA Chart for the I&T Risk Officer/Manager

Process Practice	Level of Involvement (R/A)
Develop and implement the I&T risk framework and system.	R
Develop the I&T risk management strategy and risk management plan.	R
Develop and maintain I&T risk management policies, procedures and other documents/templates.	R
Manage and execute the I&T risk program.	A
Identify I&T risk.	R
Conduct I&T risk and control assessments.	R
Assess I&T control issues; monitor, track and report.	R
Monitor I&T risk.	R
Collect and communicate threats and vulnerability intelligence.	R
Provide oversight, challenge, assurance and support to I&T risk owners.	R

3.4.3 Supporting Roles

A distinction is made between a limited number of key roles and a larger number of supporting roles. The supporting roles/structures throughout the enterprise can be found to influence I&T risk management. These roles/structures are not directly involved in I&T risk management but are relevant stakeholders in this process.

Figure 3.18 lists these relevant supporting roles and structures and provides a brief description of each and the stake each holds in supporting the I&T risk function.

Figure 3.18—COBIT Roles and Organizational Structures for I&T Risk Functions

Role	Role Description	Stake in I&T Risk Function
Board	The group of the most senior executives and/or nonexecutives of the enterprise who are accountable for the governance of the enterprise and have overall control of its resources	Provides oversight on the impact of enterprise I&T risk on the enterprise objectives[9] and makes risk decisions to protect stakeholder value at an optimal resource cost Sets the tone at the top regarding I&T risk management and awareness
Chief executive officer (CEO)	The highest-ranking officer who is in charge of the total management of the enterprise	Provides executive support and collaborates on I&T risk management decisions
Strategy (IT executive) committee	A group of senior executives appointed by the board to ensure that the board is involved in and kept informed of major IT-related matters and decisions. The committee is accountable for managing the portfolios of IT-enabled investments, IT services and IT assets, ensuring that value is delivered and I&T risk is managed. The committee is normally chaired by a board member, not by the CIO.	Being proactive in managing risk related to the portfolio of IT Investments
Chief operating officer (COO)	The most senior official of the enterprise who is accountable for the operations of the enterprise	Accountable for providing operational direction and execution, which includes operational I&T risk
Data privacy officer	The individual who oversees the development, implementation and enforcement of the data privacy and data protection policies throughout the enterprise	Accountable for providing enterprisewide compliance education Provides support and informs executive management about the risk of noncompliance with privacy laws and regulations
Chief information security officer (CISO)	The most senior officer of the enterprise who is accountable for the security of enterprise information in all forms	Accountable for providing direction and execution of information security strategies in line with I&T risk Accepts ownership of executing and monitoring the effectiveness of security controls implemented to mitigate risk
Business executive	A senior management individual accountable for the operation of a specific business unit or subsidiary. This includes key business line owners and heads of departments such as sales, marketing, human resources and manufacturing.	Accountable for providing direction and execution for business line I&T risk Accepts ownership of assigned risk and reports on mitigation progress
Chief information officer (CIO)/ chief technology officer (CTO)	The most senior official of the enterprise who is responsible for aligning IT and business strategies and accountable for planning, resourcing and managing the delivery of IT services and solutions to support enterprise objectives	Accountable for planning, resourcing and managing the IT strategy to achieve enterprise objectives
Business process owner	An individual accountable for the performance of a process in realizing its objectives, driving process improvement and approving process changes. In general, a business process owner must be at an appropriately high level in the enterprise and have authority to commit resources to process-specific risk management activities.	Consults on business processes that have an I&T risk impact Accepts ownership of assigned I&T risk and reports on mitigation progress

[9] For more information on role of the board in overseeing risk, refer to "Risk Governance: Balancing Risk and Rewards," a Blue Ribbon Commission report issued by the National Association of Corporate Directors (NACD) in 2009, *www.nacdonline.org/insights/publications.cfm?ItemNumber=675.*

Figure 3.18—COBIT Roles and Organizational Structures for I&T Risk Functions *(cont.)*		
Role	**Role Description**	**Stake in I&T Risk Function**
Chief financial officer (CFO)	The most senior official of the enterprise who is accountable for all aspects of financial management, including financial risk and controls and reliable and accurate accounts	Consults on acceptable loss exposure levels, importance of I&T risk factors and the cost of I&T risk response options
Business continuity manager	An individual who manages, designs and/or assesses an enterprise's business continuity capability to ensure that the enterprise-critical functions continue to operate following disruptive events	Responsible for managing I&T continuity risk
IT process (service) owners	The lead individual who is accountable and responsible for the IT process and/or service provided	Consults on IT processes (services) that have an I&T risk impact Accepts ownership of assigned I&T risk and reports on mitigation progress
Service manager	An individual who manages the development, implementation, evaluation and ongoing management of new and existing products and services for a specific customer, user or group of customers	Consults on products and services that have an I&T risk impact
Head of HR	The most senior official of an enterprise who is accountable for planning and policies with respect to all human resources in that enterprise	Consults about human aspects of I&T risk and actions
Privacy officer	An individual who is responsible for monitoring the risk and business impacts of privacy laws and for guiding and coordinating the implementation of policies and activities that will ensure that the privacy directives are met	Accountable for providing direction and execution for privacy-related matters that have an I&T risk impact
Project (and program) steering committee	A group of stakeholders and experts who are accountable for guidance of programs and projects, including management and monitoring of plans, allocation of resources, delivery of benefits and value, and management of program and project risk	Accountable for managing I&T-related project and program risk
Chief insurance officer	The most senior official of the enterprise who is in charge of managing insurance policies	Consulted for risk sharing/transfer
Procurement	The function that engages and manages third parties and associated processes such as contracts and overall supply chain	Consults on managing third-party I&T risk

3.5 Component: Information Flows and Items

Component C: Information Flows and Items provides guidance related to process practices. For each process practice, relevant inputs and outputs indicate the information item origin and destination—both inside and outside of COBIT—that provide input to or constitute output from the given practice (**figure 3.19**). For each governance and management objective, chapter 4 of this publication outlines detailed I&T risk-specific examples of information that is common in an I&T risk governance and management context.

Figure 3.19—Display of I&T Risk-specific Information Flows and Items Component				
C. Component: Information Flows and Items				
Governance/Management Practice	I&T Risk-specific Inputs		I&T Risk-specific Outputs	
<REF><NAME>	From	Description	Description	To
	<REF>	<TEXT>	<TEXT>	<REF>
Related Guidance (Standards, Frameworks, Compliance Requirements)	Detailed References			
<STANDARD>	<TEXT>			

Note that the practices for governance objective EDM03 *Ensured Risk Optimization* and management objective APO12 *Managed Risk* have no I&T risk-specific inputs and outputs from/to themselves, as they are already focused on risk; however, these objectives nevertheless receive other inputs and outputs from practices in other governance and management objectives.

Several I&T risk-related information types—specific to objectives EDM03 and APO12—warrant additional discussion:

- **Figure 3.20**—I&T Risk Profile
- **Figure 3.21**—I&T Risk Communication Plan
- **Figure 3.22**—I&T Risk Map
- **Figure 3.23**—I&T Risk Appetite, Tolerance and Capacity
- **Figure 3.24**—Risk Capacity, Risk Appetite and Actual Risk
- **Figure 3.25**—Key Risk Indicators
- **Figure 3.26**—Emerging I&T Risk Issues and Factors

Figure 3.20—I&T Risk Profile	
Type of Content	Specification
Description	The comprehensive view of I&T risk
Goals and quality criteria	• Source information needs to be accurate (confirmed through audit) and aggregated in the risk management application according to fixed rules. • Information is based on verifiable facts and substantiations, using the common risk view established throughout the enterprise. • The risk profile is structured as defined and the recipient confirms the relevancy of information provided. • The risk profile covers the full enterprise scope and the full risk register. However, not all information might be available and assumptions need to be made and substantiated. • The volume of information is appropriate to the recipient's needs and shall be defined during the design. • In order for stakeholders to make informed decisions, the risk profile should be understood by all relevant stakeholders. • Access to the risk profile is determined by the risk function, and is restricted as follows: ▪ Write access: risk function (based on input from contributors) ▪ Read access: all other stakeholders

Figure 3.20—I&T Risk Profile *(cont.)*	
Type of Content	**Specification**

Structure/High-level content	A risk profile is a description of the overall (identified) I&T risk to which the enterprise is exposed. An I&T risk profile consists of:
	A. Risk register
	A risk register provides detailed information on:
	• Each identified risk, such as risk owner, details of the scenario and assumptions. A risk scenario[10] is a detailed description of an IT-related risk that can lead to a business impact, when it occurs.[11] It includes elements such as:
	  ▪ Actor/threat community
	  ▪ Intent/motivation
	  ▪ Threat event
	  ▪ Assets/resource
	  ▪ Effect
	  ▪ Timing
	  ▪ Affected stakeholders
	  ▪ Causes/indicators
	  ▪ Information on the detailed scores (risk ratings) on the risk analysis
	  ▪ Detailed information on the risk response (e.g., action owner) and the risk response status (e.g., time frame for action)
	  ▪ Related projects
	  ▪ Risk tolerance level
	See Appendix C for a sample risk register template.
	B. Risk action plan
	The risk action plan provides a framework that documents the priority order in which individual risk actions shall be implemented and how they shall be implemented. An action plan should be discussed with appropriate stakeholders and clearly document:
	• Risk scenarios that will be mitigated by the identified actions
	• Root cause of the scenario (root cause analysis [RCA])
	• The reasons for selection of action options based on the control evaluation criteria
	• Those who are accountable for approving the plan and those who are responsible for implementing the plan
	• Proposed actions
	• Resource requirements, including contingencies
	• Performance measures and constraints
	• Cost vs. risk reduction benefit
	• Reporting and monitoring requirements
	• Timing and schedule
	C. Loss events (historical and current)
	A loss event is a risk event that resulted in a loss. For example, an error in systems processing/timing of trades can result in a loss or negative position. It should include information on root cause and actual cost of the event, ideally subdivided by the enterprise's standard loss forms.
	D. Risk factors
	A risk factor is a condition that can influence the frequency and impact and, ultimately, the business impact of IT-related events/scenarios. Risk factors can also be interpreted as causal factors of the scenario that is materializing, due to vulnerabilities or weaknesses. These factors include:
	• External context

[10] See ISACA, *Risk IT Practitioner Guide, 2nd Edition*, USA, 2020, *www.isaca.org/bookstore/bookstore-risk-print/ritpg2*, for guidance on risk scenarios.

[11] A list of generic risk scenarios and example scenarios is included in Appendix A. These scenarios were leveraged from ISACA's *Risk IT Practitioner Guide, 2nd Edition*, 2020. The scenarios are grouped into 18 risk scenario categories. Appendix B of this publication displays a mapping of these risk scenario categories to COBIT 2019 governance and management objectives.

Figure 3.20—I&T Risk Profile *(cont.)*	
Type of Content	**Specification**
Structure/High-level content *(cont.)*	• Internal context • Risk management capabilities • IT-related capabilities **E. Independent assessment findings** This is an audit report that contains findings of assessments that should be taken into consideration during risk identification and analysis activities.

Figure 3.21—Risk Communication Plan	
Type of Content	**Specification**
Description	A risk communication plan defines the frequency of reporting, types of information about I&T risk and recipients of that information. The main purpose of the plan is to reduce the overload of irrelevant information (avoiding the likelihood of "risk noise").
Goals and quality criteria	The risk communication plan: • Should accurately define the type, frequency and recipient of risk communications • Ensure that information is based on the enterprise's risk culture • Is devised based on input from both the risk function and business process owners, making the devised risk communication plan appropriate • Shall be aligned to the needs of the recipients • Shall cover the end-to-end enterprise structure as well external stakeholders • Shall not be older than one year • Should contain communication frequencies, types and recipients • Ensures that information is always presented according to the predefined formats/templates • Should be understandable for the target audience • Shall be available at the required frequency to key stakeholders. Information to the risk function should be available 24/7. • Indicates that the access to this information item is determined by the risk function, and is restricted as follows: ▪ Write access: risk function ▪ Read access: all other stakeholders
Structure/High-level content	The risk communication plan contains the following subparts: • Frequency/timing of communication • Type of communication • Target audience and distribution

Figure 3.22—I&T Risk Map (see also Appendix E)	
Type of Content	**Specification**
Description	The risk map is a common, very easy and intuitive technique to present risk. It plots risk on a two-dimensional diagram, with frequency and impact being the two dimensions. The risk map representation is powerful and provides an immediate and complete view on risk and apparent areas for action. Furthermore, a risk map allows defining color zones to indicate graphically risk appetite bands of significance. A practical consideration is to focus on the 10 or 20 most important risk items, or split them up into categories, as the map would otherwise not be readable.

Figure 3.22—I&T Risk Map *(cont.)* (see also Appendix E)	
Type of Content	**Specification**
Goals and quality criteria	The I&T risk map *(cont.)*: • Should accurately define the risk status of all risk items • Ensures that information is collected/confirmed by multiple sources in the enterprise and external to the enterprise • Shall be organized according risk categories as per the agreed risk taxonomy to ensure adequate comparison possibilities • Shall cover the end-to-end enterprise risk spectrum • Shall present current status • Should contain high-level information. The volume of source information should be appropriate to support the indication in the risk map. • Should be understandable for the target audience • Shall be available 24/7
Structure/High-level content	In creating the risk map, it is critical to determine the most important risk items to plot on the map or determine the categories to show.

Figure 3.23—I&T Risk Appetite, Tolerance and Capacity	
Type of Content	**Specification**
Description	Risk appetite is the broad-based amount of enterprise I&T risk that an enterprise is willing to accept in pursuit of its mission (or vision). Risk tolerance is the acceptable level of variation that management is willing to allow for any particular risk as it pursues the enterprise's objectives. Risk capacity is the objective magnitude or amount of loss that an enterprise can tolerate without risking its continued existence.
Goals and quality criteria	Define the enterprise's criteria to manage and respond to risk.
Structure/High-level content	See the discussion in the following section.

About I&T Risk Appetite, Risk Tolerance and Risk Capacity

In formulating strategies and/or operating plans, an enterprise must decide to take on some level of risk to achieve its objectives. An amount or magnitude of risk is generally expressed as risk appetite and risk tolerance. Although these terms are used frequently, the potential for misunderstanding is high. Some people use the concepts interchangeably, others see a clear difference. ISACA's Risk IT framework definitions are compatible with the Committee of Sponsoring Organizations of the Treadway Commission (COSO) ERM16 and the International Organization for Standardization (ISO) 31000 definitions:

- **Risk appetite**—The broad-based amount of risk an enterprise or other entity is willing to accept in pursuit of its mission (or vision)
- **Risk tolerance**—The acceptable range relative to the achievement of a given objective (best when quantified in terms of in the same unit measure as the related objective)

Risk Appetite

Risk appetite reflects the amount of risk an entity is prepared to accept in order to achieve its objectives. When considering risk appetite levels for the enterprise, three major factors are important:

- The objective capacity of the enterprise to absorb loss, for example, financial loss or damage to reputation

- The management culture or predisposition toward risk-taking, for example, cautious or aggressive. What amount or magnitude of loss will the enterprise accept to pursue its strategy or objectives?
- The nature of the business and the type of risk involved, for example, the failure of a conveyor belt in a candy factory vs. the failure of a flight-control system on a commercial airliner

Risk appetite is different in each enterprise—there is no absolute norm or standard of what constitutes acceptable and unacceptable risk.

Statements of risk appetite are often broad, and tend to speak of risk hypothetically or generally—such as, "the enterprise will not accept the risk of noncompliance" or "the organization will not accept fraud risk"—rather than express risk concretely in quantifiable terms. Although such representations of risk appetite are common, they are very difficult to cascade down through the organization as management directives.

Absolute prohibitions on risk are impossible to maintain and therefore impractical. Under such a prohibition against risk, every control deficiency would be fixed and every business endeavor with risk would be declined. In practice, this approach is not a productive or efficient use of resources. Instead, enterprises should attempt to determine a loss amount that is acceptable and manage to that amount. An example of a practical, concrete, quantified risk appetite statement might be:

Although the enterprise desires to have no appetite for I&T risk, it recognizes that this is impractical in the achievement of its objectives. Therefore, the enterprise will remediate loss scenarios whereby aggregate losses of US $1 million or more are at risk.

Large enterprises may find it useful to have a version of this statement for each line of business. An enterprisewide appetite statement should reflect (or aggregate) all the line-of-business statements.

Every enterprise must define its own risk appetite levels and review them on a regular basis. This definition of risk appetite should align with the overall risk culture that the enterprise wants to express (i.e., ranging from very risk-averse to risk-taking/opportunity-seeking). Although there is no universal right or wrong, risk appetite needs to be defined, well understood and communicated. Risk appetite and risk tolerance should be applied not only to risk assessments but also to all I&T-related decision making.

Risk Tolerance

Risk tolerance reflects a range of acceptable deviation from the level set by the risk appetite and business objectives, for example:

Standards require projects to be completed within estimated budgets and time frames, but overruns of 10 percent of budget or 20 percent of time are tolerated.

Guidance Regarding Risk Appetite and Risk Tolerance

The following guidance applies to risk appetite and risk tolerance:

- Risk appetite and tolerance are defined at the enterprise level, reviewed and/or influenced by the board of directors, and reflected in strategy and policies set by executives. At lower (tactical) levels of the enterprise—or perhaps within certain enterprise entities or subsidiaries—exceptions can be tolerated (or different thresholds defined), as long as overall exposure at the enterprise level does not exceed the determined risk appetite. Any business initiative includes a risk component, so management should have discretion to pursue new opportunities in the context of risk. Enterprises that have a conservative risk appetite and tolerance policies could lack the agility or innovation to exploit new business opportunities. Conversely, risk appetite and tolerance policies can be dictated by legal, regulatory or industry requirements, and it may be appropriate to have no risk tolerance for failure to meet such mandates.

- Risk appetite and tolerances are defined, reviewed and updated periodically (as determined by the enterprise), and clearly communicated to all stakeholders. Risk exception processes are clearly defined and communicated.
- New market conditions, changing risk landscape, revised strategy and many other factors require the enterprise to reassess its risk portfolio and reconfirm its risk appetite at regular intervals, triggering risk policy reviews. In this respect, the enterprise should understand that risk management can provide value to the enterprise by allowing it to pursue risk-inclusive strategies and optimize allocation of resources.
- The costs of risk response or the business impact of risk may exceed the capabilities/resources of an enterprise, thus forcing higher tolerance for one or more risk conditions. For example, if a regulation says that sensitive data at rest must be encrypted—but no feasible encryption solution exists, or the cost of implementing a solution is prohibitive—the enterprise may choose to accept the risk associated with regulatory noncompliance, which is an enterprise-appropriate decision informed by factual data.

Risk Capacity

The term risk capacity is sometimes used in discussions of risk appetite. Risk capacity is usually defined as the objective magnitude or amount of loss that an enterprise can tolerate without risking its continued existence. As such, it differs from risk appetite, which generally reflects a board or management decision regarding how much risk is desirable, as illustrated in **figure 3.24**.

Figure 3.24—Risk Capacity, Risk Appetite and Actual Risk

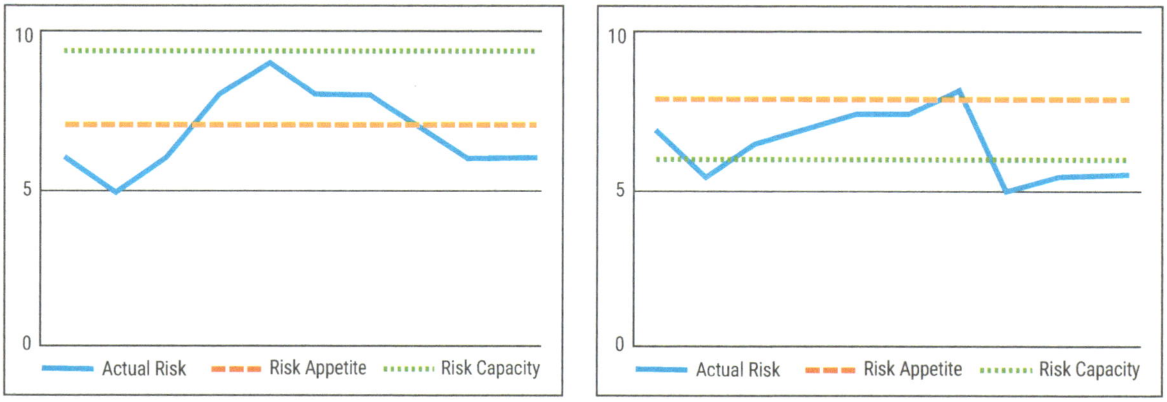

Source: ISACA, *COBIT® 5 for Risk*, USA, 2013, figure 68, *www.isaca.org/bookstore/cobit-5/wcb5rk*

- The left diagram shows a relatively sustainable situation in which risk appetite is lower than risk capacity, and actual risk exceeds risk appetite in several situations, but always remains below the risk capacity.
- The right diagram shows a rather unsustainable situation, where risk appetite is defined by management at a level beyond risk capacity. Management is prepared to accept risk well over the objective capacity to absorb loss. As a result, actual risk routinely exceeds risk capacity, despite remaining below the risk appetite level most of the time.

Key Risk Indicators

Key risk indicators (KRIs) are metrics capable of showing that the enterprise is subject to—or has a high probability of being subject to—a risk that exceeds the defined risk appetite or tolerance (**figure 3.25**). As their name implies, they are only indicators of risk and not direct measures of risk. It is important not to confuse risk measurement (and the corresponding assignment of risk ratings) with KRIs. They are specific to each enterprise, and their selection depends on many parameters in the internal and external environment, including the size and complexity of the enterprise, regulatory context (i.e., whether it operates in a highly regulated market), and strategy focus.

Figure 3.25—Key Rey Indicators

Type of Content	Specification
Description	Key risk indicators (KRIs) are metrics capable of showing that the enterprise is subject to—or has a high probability of being subject to—a risk that exceeds the defined risk appetite or tolerance.
Goals and quality criteria	KRIs: • Need to be accurate and should accurately articulate thresholds • Should be based on objective information—measurable components • Should indicate that people are convinced of the correct definition and contribution to the insight in the risk exposure of the enterprise • Should indicate that people are convinced of the contribution to the insight of the exposure of the enterprise • Should contain all the key elements that have an influence on the risk that the indicator is quantifying • Should consist of a composition and goal/meaning that is clear to the target audience
Structure/High-level content	The risk indicators contain metrics and an explanation of the goal/measurement.

Identification of KRIs should take into account the following steps (among others):

1. **Consider stakeholder needs when developing indicators of risk**. KRIs should be identified for relevant stakeholders based on information needs. Involving the right stakeholders in the selection of risk indicators also ensures greater buy-in and ownership.
2. **Iterate and improve indicators over time**. Take a balanced approach in selecting indicators that are forward-looking, or leading, and backward-looking, or lagging.

Leading indicators include various data attributes, information or capabilities that are in place to prevent events from occurring. Leading indicators may have upper and lower limits to help an enterprise understand when a condition requires attention before the risk is realized.

Lagging indicators include various data attributes, information or capabilities that are measured after an event or condition occurs, for example, meeting a performance target or service level availability goal. Analyzing root causes from realized risk, failed controls or processes, and missed targets over time can help enterprises develop new indicators, trends or correlating conditions to gain insights.

An enterprise may develop an extensive set of metrics to serve as risk indicators; however, it is not generally feasible to maintain a large set of KRIs. By definition, *key* indicators are differentiated as being highly relevant and possessing a high probability of predicting or indicating risk outcomes.

The selection of the appropriate KRIs provides the following benefits to the enterprise:

- **Early warning, forward-looking signals** to the enterprise that a risk may be soon realized, enabling a proactive response before the risk becomes a loss
- **Backward-looking historical context** on risk that has been realized, further informing future risk responses, driving improvement, and supporting documentation and analysis of trends
- **Feedback on risk appetite and tolerances** to facilitate improvements in risk management strategy and processes, and optimize risk governance and oversight

Common challenges or pitfalls associated with KRIs include:

- Undefined measurement objectives, absence of stated/expected outcomes, or lack of definitive questions that can be answered with data from KRIs
- Rote collection of data that are easy to obtain or already on hand, as opposed to data that correlate significantly to specific risk or risk types
- Lack of a clear logical relationship between KRIs and specific risk or business objectives
- Excess of metrics without a clear measurement objective or purpose
- Cumbersome processes of aggregation
- Excess complexity in synthesizing and/or interpreting results from KRIs at an enterprise level

Because the internal and external environments are constantly changing, the risk environment is also highly dynamic, and the set of KRIs needs to be changed over time. Each KRI should be clearly related to the risk appetite and tolerance, so that trigger levels can be defined in support of appropriate, timely action (**figure 3.26**).

Figure 3.26—Emerging I&T Risk Issues and Factors

Type of Content	Specification
Description	Emerging I&T risk issues and factors are information on upcoming or likely combinations of control, value and threat conditions that impose a noteworthy level of future I&T risk.
Goals and quality criteria	• Source information needs to be accurate and must be processed, analyzed and presented in the enterprise's standard format. • Information is based on verifiable facts and substantiations, using the coming risk view established throughout the enterprise. • Source information is collected from competent and recognized sources. • Information should be current (no older than a month). • The information will be new to many stakeholders so it needs to be easy to comprehend.
Structure/High-level content	• The content is complete and of sufficient depth and breadth. • The content is concisely presented.

See Appendix D for I&T risk reporting examples including a dashboard, graphs and other metrics.

3.6 Component: People, Skills and Competencies

In *COBIT 2019 Framework: Governance and Management Objectives*, the people, skills and competencies governance component identify the human resources and skills required to achieve the governance or management objective.

COBIT 2019 based this guidance on the *Skills Framework for the Information Age* (SFIA®), version 6.[12] All skills are described in detail in the SFIA framework. (Because there is no change from the COBIT core model, references to SFIA are not detailed in *COBIT Focus Area: Information & Technology Risk*.)

[12] SFIA Foundation, *Skills Framework for the Information Age*, version 6, United Kingdom, 2015, *https://sfia-online.org/en/legacy-sfia/sfia-6*

The purpose of this section is to identify and describe all skill sets and competencies that are required to build and sustain effective and efficient risk management in an enterprise. In other words, it lists all skill sets and competencies needed to support the risk function (**figure 3.27**).

Figure 3.27—Risk Management Skill Sets and Competencies

Skill Set and Competencies	Description
Leadership skills	Risk management often involves many different stakeholders with different opinions and values that must be navigated to drive effective business outcomes, requiring effective risk management leadership. Leaderships skills include proactive leadership that sets clear direction that is aligned to the business outcomes and determination to ensure that the implemented policies deliver the effective disposition of risk. Leadership skills also require the ability to effectively work with all stakeholders to demonstrate effective escalation and communication. Finally, risk needs to be managed in a cost-effective manner.
Analytical capability	The increasing complexity of business and the need to comply with regulations for each vertical industry require that risk analysts have the capabilities to break down risk into risk factors that may prevent the achievement of goals and to assess those risk factors. This increasing complexity of business, legislative and regulatory requirements demands analytical skills to decipher them. The ability to analyze the entire risk as pieces to see where the fault lies and then to put the pieces together in a useful and understandable manner requires the risk analyst to approach the topic methodically and with a structured mindset.
Critical thinking	Among the skills required for effective risk management is the ability to: • Make professional judgments about the value of additional information and determine whether a sufficient level of analysis has occurred • Document and qualify assumptions and articulate risk scenarios
Interpersonal capabilities	A key attribute of risk professionals is their ability to obtain information that is timely and accurate and communicate with stakeholders who have different backgrounds and objectives. The risk professional uses nontechnical language effectively, so that the message is meaningful to all stakeholders, and demonstrates an essential understanding of business goals and priorities.
Communication	Risk needs to be communicated to stakeholders who have different backgrounds and objectives. The risk manager or analyst should have the capability to communicate risk, risk factors and the associated loss exposure in the context, language and priority of the relevant stakeholder. The risk manager or analyst is capable of engaging all stakeholders in a meaningful way, using consistent nomenclature and providing examples for context.
Influencing	Risk professionals require well-developed persuasion skills to help with adoption of risk practices across the enterprise and demonstrate value to stakeholders.
Lateral thinking (thinking outside the box)	Risk needs to be approached differently depending on the type of risk. Ideas and techniques from other disciplines should be leveraged.
Technical understanding	The level of technical skill and competencies depends on the role within the risk function. Understanding the vulnerabilities of IT systems and the threats that exploit them requires a basic understanding of the components comprising IT systems and how these components are connected to each other physically and logically.
Organizational and business awareness	To enable the enterprise to effectively plan, communicate and execute its risk management processes, the organizational points of contact, business units, goals, employee roles and responsibilities, and escalation paths must be documented and kept up to date.
Risk expertise	Expertise in threat sources, threat scenarios, vulnerabilities and impact on the business is critical to success. This skill refers to an understanding of the basic nature and composition of risk as well as ongoing improvement to keep pace with the dynamic nature of threats, vulnerabilities and impacts in the modern business environment.
Training and coaching	Risk management is a part of all roles in the enterprise. To confirm appropriate levels of risk expertise and practice, enterprises require capabilities in effective training and coaching of stakeholders.

Figures 3.28 through **3.30** further outline the skills and competencies of three specific risk roles with risk management responsibilities:

Figure 3.28—Chief Risk Officer

A **CRO** is responsible for championing initiatives to mitigate business risk that can put an enterprise's profitability and productivity at risk. The role also oversees the enterprise risk management (ERM) program and is responsible for implementing policies and procedures to minimize or manage operational risk.

Requirement	Description
Experience, Education and Qualifications	
Experience	• More than 10 years of experience of managing technical/risk/compliance team members • Ability to evaluate risk and prioritize work
Education	Master's degree in economics, business, computer science, or statistics
Typical qualifications and certifications	Certified in Risk and Systems Controls™ (CRISC™), Certified Regulatory Compliance Manager (CRCM), National Association of Federal Credit Unions (NAFCU) Certified Compliance Officer (NCCO)
Knowledge, Technical Skills and Behavioral Skills	
Knowledge	• Have deep knowledge of risk methodologies and practices • Have knowledge of strategic planning techniques • Have knowledge of process management and people management
Technical skills	• A background in finance and technology • Strong managerial and analytical skills • Strong communication skills
Behavioral skills	• Strong leadership skills and the ability to effectively communicate with senior business leaders • Ability to delegate and oversee staff members • Ability to think strategically and critically

Figure 3.29—Risk Manager

Risk managers are responsible for the successful implementation and monitoring of the risk strategy and framework. They engage with stakeholders to ensure that risk management processes are understood, resourced and implemented, and support business goals. Risk outcomes are reported to the CRO for incorporation into overall risk profiles and risk issues.

The role works with business management to ensure that the overall I&T risk function effectively supports strategic goals. The risk manager collaborates with audit/business segment/corporate risk to address issues with plausible action plans and target dates. This role acts as the central point for receipt and distribution of important risk information for IT and reciprocates the flow of information back to corporate and business unit policies and procedures. The role must be aware of and keep abreast of technology risk associated with the enterprise. The role may or may not have managerial responsibility.

This figure describes the typical experience, education and qualifications for this specific role. These should not be considered strict requirements, but guidance that can be used as input, for example, when detailing job descriptions.

Requirement	**Description**
Experience, Education and Qualifications	
Experience	• Adequate experience in managing and governing business risk and/or operations • Experience in communication of risk to executive management and/or board
Education	Degree in management information systems with experience in I&T, finance, economics, business or engineering
Typical qualifications and certifications	Certified Information Systems Auditor® (CISA®), Certified Information Security Manager® (CISM®), Certified in Risk and Systems Controls™ (CRISC™), Certified Information Systems Security Professional (CISSP), Certified Public Accountant (CPA)

Figure 3.29—Risk Manager *(cont.)*	
Knowledge, Technical Skills and Behavioral Skills	
Requirement	**Description**
Knowledge	• A deep knowledge of the enterprise and the I&T systems that support the business functions as well as awareness of the contextual factors that influence them • Solid knowledge of risk methodologies, commonly used risk standards and risk best practices
Technical skills	Knowledge of the technical side of I&T systems supporting the business functions
Behavioral skills	• Leadership • Communication • Influencing • Patience • Escalation

Figure 3.30—Risk Analyst	
The risk analyst is responsible for: • Executing the overall risk assessment process in the enterprise • Identifying and analyzing the areas of potential risk that are threatening assets and the achievement of the organizational objectives • Providing specific evaluation of risk scenarios by considering the business and the technical perspective • Reporting findings to the risk manager or CRO This figure describes the typical experience, education and qualifications for this specific role. These should not be considered strict requirements, but guidance that can be used as input, for example, when detailing job descriptions.	
Experience, Education and Qualifications	
Requirement	**Description**
Experience	• Adequate relevant experience in business administration or IT • An authoritative knowledge in systems architecture, infrastructure, security and applications
Education	• Bachelor's degree in financial analysis, IT, engineering, systems analysis • Master's degree in a related discipline, such as mathematics, statistics
Typical qualifications and certifications	CISM, CRISC, CISSP, Factor Analysis of Information Risk (FAIR)
Knowledge, Technical Skills and Behavioral Skills	
Knowledge	• Knowledge of risk methodologies, commonly used risk standards, risk good practices, and quantitative and qualitative risk analysis • Consistent knowledge of business processes and their relationship to technology
Technical skills	• Profound IT and business functioning understanding and a profound understanding of IT domains, threats, assets • Analytical capability, preferably with knowledge of statistical analysis and probabilities
Behavioral skills	• Communication skills • Presentation skills • Peer reviews • Decision making • Ability to delegate work to technical experts

3.7 Component: Principles, Policies and Procedures

3.7.1 Principles

Principles, policies and procedures document and communicate the enterprise rules that support governance objectives and enterprise values. To supplement the guidance on principles, policies and procedures in *COBIT® 2019 Framework: Governance and Management Objectives*, this section elaborates detailed I&T risk-specific principles and policies.

Any I&T-related risk that jeopardizes the organization's business or mission should be managed from the perspective of overall enterprise objectives, and thus falls under the guiding principles of the Risk IT framework (**figure 3.31**):

- Connect management of I&T-related risk to business or mission objectives.
- Align the management of I&T-related business or mission risk with ERM when possible.
- Balance the costs and benefits of managing I&T-related risk.
- Promote ethical and open communication of all I&T-related risk.
- Establish the tone at the top while defining and enforcing personal accountability for operating within acceptable and well-defined tolerance levels.
- Use a consistent approach, integrated into daily activities, that is standard, repeatable and aligned to strategy.

Figure 3.31—Principles of Risk Governance and Management

Source: ISACA, *COBIT® 5 for Risk*, USA, 2013, figure 15, *www.isaca.org/bookstore/cobit-5/wcb5rk*

Connect to Enterprise Business or Mission

Effective principles for enterprise governance and management of I&T-related risk always connect to business or mission objectives:

- I&T-related risk, including cyberrisk, is treated as a business risk, not as a separate type of risk, and the approach to management is comprehensive and cross-functional.

- Governance of I&T-related risk contributes to business or mission outcomes. I&T support the achievement of business objectives, and any associated risk is expressed in terms of the impact and probability it may have on business objectives or strategy. Analysis of I&T-related risk considers the connection between business processes and supporting I&T assets, applications or infrastructure, and/or third-party dependencies.
- I&T-related risk management, including practices in information security and cybersecurity, all strive to advance the business or mission, rather than limit or inhibit it.

Align With Enterprise Risk Management

Effective enterprise governance of I&T-related risk aligns its management with overall ERM:

- Business or mission objectives and risk appetite are clearly defined.
- Enterprise decision-making processes consider the full range of potential consequences and opportunities from I&T-related risk.
- The defined and stated risk appetite reflects the ERM policy and tone at the top and influences the culture of the enterprise.
- I&T-related risk assessment is coordinated and consolidated across the enterprise (including, for example, across information security and cybersecurity).

Balance Costs and Benefits

Effective enterprise governance of I&T-related risk balances its costs and benefits:

- I&T-related risk is prioritized and addressed in line with risk appetite and risk tolerance.
- Risk responses are implemented based on cost/benefit analysis, analysis of alternatives and prioritization of risk that has the greatest potential impact on enterprise objectives.
- Existing controls and risk response actions are leveraged to address risk as efficiently as possible.

Promote Ethical and Open Communication

Effective management of I&T-related risk promotes ethical and open communication:

- Open, accurate, timely and transparent information on I&T-related risk is freely exchanged and informs risk-related decisions.
- Risk culture and risk management methods are integrated across the enterprise.
- Technical findings are translated into relevant and understandable business and financial terms.
- Information about an incident and the associated response is communicated openly to stakeholders, government and/or regulatory authorities, customers, and (as necessary) the public.

Establish Tone at the Top and Accountability

Effective management of I&T-related risk establishes an engaged tone from the top while defining and enforcing personal accountability for operating within acceptable and well-defined tolerance levels:

- Business owners, the board of directors and executive leadership are engaged in risk management.
- There is clear accountability and assignment of risk ownership.
- Risk assumptions are understood and supported by appropriate business leaders and are clearly stated in documentation of risk appetite, tolerances, culture, policies and guidelines for enforcement.

- Risk management performance is measured and integrated into the performance management of those accountable and responsible.
- Risk-aware culture and personal responsibility are promoted.
- Risk-informed decisions are made at the right level in the organization, by authorized individuals, in line with tolerances.
- Risk management practices are appropriately prioritized and embedded into enterprise decision making.

Use a Consistent Approach Aligned to Strategy

Effective management of I&T risk promotes continuous improvement and is part of daily activities:

- The dynamic nature of risk requires the enterprise to prepare by giving advance consideration to changes in:
  - The organization itself (mergers and acquisitions)
  - The risk landscape
  - Applicable laws and regulations
  - I&T, as they evolve
  - The industry at large
- Risk assessment methods, scales of measurement and criteria are consistent across the enterprise, especially as applied to:
  - Identification of key processes and associated risk
  - Identification of impacts on objectives
  - Identification of triggers that indicate when risk is out of tolerance or when an update of the framework or components in the framework are required, etc.
  - Monitoring and testing of operating controls
  - Actions to prevent risk from materializing
  - Risk response (if adverse events occur)
  - Identification and, to the extent possible, mitigation of assessor bias in the quantitative risk measurement process

3.7.2 Policies

Supporting Policies

Policies provide more detailed guidance on how to put principles into practice and how they will influence decision making. Not all relevant policies are written and owned by the IT risk function. A number of risk-related policies are described in this publication, and the policy driver within the enterprise is specified.

The core IT risk policy is shown in **figure 3.32**, which also provides examples of operational-level policies that should be considered for complete I&T risk management. Depending on the size and nature of the enterprise, these may be policies in their own right or sections within existing policies. For operational efficiency, it is necessary to keep these policies in sync with the risk policy.

Figure 3.32—Risk Policy Examples	
Policies	**Description**
Core IT risk policy	Defines, at strategic, tactical and operational levels, how the risk of an enterprise needs to be governed and managed pursuant to its business objectives. This policy translates enterprise governance into risk governance principles and policy and elaborates risk management activities.
Information security policy	Sets behavioral guidelines in protecting corporate information and the associated systems and infrastructure. The business requirements regarding security and storage are more dynamic than IT risk management, so, for effectiveness, their governance needs to be handled separately from the governance of IT risk. However, for operational efficiency, it is necessary to keep the information security policy in sync with the IT risk policy.
Crisis management policy	As with IT security, network management and data security, IT crisis management is one of the operational-level policies that needs to be considered for complete IT risk management. It sets the guidelines on how to act in situations of crisis and details the sequence in which to deal with each of the identified (key) areas of risk.
Third-party IT service delivery management policy	Sets guidelines for managing the risk related to third-party services. It sets out a framework of expectations in behavior and security precautions taken by third-party service providers to manage the risk related to service provision.
Business continuity policy	Contains management commitment and view on: • Business impact analysis (BIA) • Business contingency plans with trusted recovery • Recovery requirements for critical systems • Defined thresholds and triggers for contingencies • How to handle escalation of incidents • Disaster recovery plan (DRP) • Training and testing
Program/project management policy	Deals with managing risk linked to projects and programs. It details management position and expectation regarding program and project management. Moreover, it handles accountability goals and objectives regarding performance, budget, risk analysis, reporting and mitigating adverse events during the execution of programs and projects.
Human resources (HR) policies	Detail what employees can expect from the enterprise and what the enterprise expects from employees. They provide detailed acceptable and unacceptable behavior by employees, and, in doing so, manage the risk that is linked to human behavior.
Fraud risk policy	Is concerned with protecting the enterprise brand, reputation and assets from loss and/or damage resulting from incidents of fraud and/or misconduct. The policy provides guidance to all employees on reporting any suspicious activity and ways to handle sensitive information and evidence. It helps to build an antifraud culture and raise awareness of the risk.
Compliance policy	Explains the assessment process regarding compliance with regulatory, contractual and internal requirements. It lists roles and responsibilities for the different activities in the process and provides guidance on metrics to be used to measure compliance.
Ethics policy	Defines the essentials of how people within an enterprise will interact with one another, as well as how they will interact with any customers or clients they serve.
Quality management policy	Details management's vision on the quality objectives of the enterprise, the acceptable level of quality and the duties of specific departments to ensure quality.
Service management policy	Provides direction and guidance to ensure the effective management and implementation of all IT services to meet business and customer requirements, within a framework of performance measurement. It also deals with the management of risk related to IT services. Detailed guidance on service management and the optimization of risk related to services is included in the ITIL framework.
Change management policy	Communicates management's intent that changes to enterprise IT be managed and implemented in a way that minimizes risk and impact to the stakeholders. The policy contains information on the assets in the scope and the established standard change management process.

Figure 3.32—Risk Policy Examples *(cont.)*	
Policies	**Description**
Delegation of authority policy	Outlines: • The authority that the board strictly retains for itself • The general principles of delegation and authority • A schedule of the delegation of authority (including clear boundaries) • A clear definition of the organizational structures to which the board delegates its authority
Whistleblower policy	Should: • Encourage employees to raise concerns and questions • Provide avenues for employees to raise concerns in full confidence • Ensure that employees will receive a response to raised concerns and be able to escalate a concern if they are not satisfied with the response • Reassure the employees that they are protected when they raise issues and should not be afraid of reprisals
Internal control policy	The purpose is to: • Communicate management's internal control objectives • Establish standards for the design and operation of the enterprise system of internal controls to reduce the exposure to all risk faced by the enterprise
Intellectual property (IP) policy	The purpose is to ensure that all risk related to the use, ownership, sale and distribution of the outputs of I&T-related creative endeavors by employees of an enterprise, such as software development, is detailed in an appropriate way from the start of any endeavor.
Data privacy policy	A statement or a document that discloses the ways that a party gathers, uses, discloses and manages personal data. Personal information can be anything that can be used to identify an individual, including, but not limited to, name, address, date of birth, marital status, contact information, ID issue and expiry date, financial records, credit information, medical history, travel destination, and intentions to acquire goods and services. The policy defines how an enterprise collects, stores and releases the personal information it collects. The policy informs the client of the specific information that is collected and whether it is kept confidential, shared with partners, or sold to other firms or enterprises. Furthermore, the policy ensures compliance with relevant legislation related to data protection.

Core Risk Policy

Figure 3.33 is an example table of contents for a core risk policy. Details about each of the items listed are provided after the figure, often noting references to other sections in this publication.

Figure 3.33—Risk Policy Table of Contents Example
Table of Contents
1. Scope
2. Validity
3. Management Commitment and Accountability
4. Risk Governance
4.1 Principles
4.2 Evaluate
4.2.1 Stakeholder Needs
4.2.2 Drivers and Goals
4.3 Direct
4.3.1 Enterprise
4.3.2 Roles and Responsibilities
4.3.3 Objectives
4.4 Monitor
4.4.1 Metrics
4.4.2 Communication
5. Risk Management Framework

1. Scope

The following text provides an example of a scope declaration in a risk policy.

> *A risk policy comprises governing and managing all externally and internally driven as well as strategic, delivery and operational risk. It helps to demonstrate a good practice of risk governance and management, earn competitive advantage and support business objectives. Moreover, risk governance and management are also viewed as essential components of corporate governance, strongly supported by the organization's internal control system and value charter.*

To be effective, risk management must be embedded in the normal process and form part of the daily management practice. This will lead to sound, risk-based decision making; establish a risk-aware culture among all employees at all levels; and provide the necessary assurance to stakeholders.

The purpose of this policy is to deploy a risk management process within corporate units in accordance with the IT strategic plan, which provides the vision, objectives and principles of IT to be applied throughout the enterprise and the basic guidelines on how to apply them in practice. The objective of this policy is to build a sustainable &IT risk framework that supports managing I&T risk in a cost-effective and pragmatic way, while complying with requirements.

2. Validity

Figure 3.34 describes the three aspects of validity that should be clearly identified in a risk policy.

Figure 3.34—Validity Aspects to be Identified in a Risk Policy

Validity Aspect	Description
Applicability	States to whom in the enterprise that the policy is applicable
Update and revalidation	States the frequency of update of the policy and the persons from the enterprise who will revalidate the updated policy
Distribution	States the distribution list of the risk policy. A difference can be made between the people to whom the policy is explicitly communicated and those to whom the policy is available. It is the responsibility of top management to make sure that the distribution happens appropriately, with a minimum distribution to which the policy is applicable.

3. Management Commitment and Accountability

Probably the most important influencing factor in effective risk management is the demonstration of executive management's support for the risk management program. This means that reasonable consideration needs to be given and appropriate action should be taken on viable proposals or recommendations.

For that reason, holding management at all levels of the hierarchy accountable for the fair and consistent application of the risk management practices is a key influencing factor. This commitment and accountability should be clearly communicated and detailed in the risk policy.

4. Risk Governance

As noted in **figure 3.33**, the risk governance section of the risk policy consists of several components. The following outline describes each component in **figure 3.33**, with references to other content in this publication as applicable.

4.1 **Principles**—Risk governance principles are listed in section 3.7.1.

4.2 **Evaluate**—Balanced, agreed-on enterprise objectives to be achieved must be determined.

4.2.1 **Stakeholder Needs**—Stakeholders for risk management need to be defined. **Figure 3.4** identifies all potential stakeholders and their interest. Based on that information, this section in the policy can be completed.

4.2.2 **Drivers and Goals**—The goals of risk management should be defined based on the needs of stakeholders listed in the preceding item (item 4.2.1) and on the enterprise goals and IT-related goals.

4.3 **Direct**—Direction of risk management is expressed in activities, accountability, and roles and responsibilities by objectives.

4.3.1 **Enterprise**—The organizational structure(s) put in place to manage risk should be described; the organizational structures governance component (section 3.4) can be used for that purpose. That section contains a much more detailed description of these organizational structures, their key practices and how they can be implemented, and can be used to define this part of the risk policy.

4.3.2 **Roles and Responsibilities**—These should be established where key responsibilities are defined. Overall accountability should be clearly assigned to the board.

4.3.3 **Objectives**—The key objectives of risk management should be defined and aligned with enterprise goals. The goals should be specific, measurable, attainable, realistic and timely (SMART) and should address efficiency and effectiveness of risk management and related processes.

4.4 **Monitor**—Risk activity should be monitored based on standard risk reporting that allows decision making.

4.4.1 **Metrics**—The appropriate metrics should be defined for the measurement of risk management and related processes.

4.4.2 **Communication**—Risk capabilities, risk status and risk profile should be maintained and communicated to all relevant stakeholders. Various communication mechanisms are identified in this guide, for example:

- Relevant risk management processes
- Appropriate information items. The most important here is the risk profile; the risk policy could, for example, identify the key components the risk profile should contain and how often these should be made available.

5. Risk Management Framework

The risk management framework defines—at a high level—all governance components the enterprise will put in place for its risk management efforts.

3.8 Component: Culture, Ethics and Behavior

3.8.1 Desirable Behaviors

This section identifies relevant behaviors and culture elements that are required to build and sustain effective and efficient risk management in an enterprise and that contribute to establishing and maintaining a risk-aware culture at all levels of the enterprise. In other words, it lists relevant behaviors that support the risk function. The desirable behaviors are categorized according to three levels within the enterprise:

- General (enterprisewide)
- Risk professionals
- Management

The goal or key objective, suitable criteria, or desired outcomes are listed for each behavior.

In **figure 3.35**, a detailed description can be found of how culture aspects such as communication, rules, incentives, rewards and raising awareness can influence these behaviors.

Figure 3.35—Relevant Behavior for Risk Governance and Management

Behavior	Key Objective/Suitable Criteria/Outcome
General (Enterprisewide) Behavior	
Has a risk- and compliance-aware culture throughout, including the proactive identification and escalation of risk	Must define a risk management approach and risk appetite. Zero tolerance of noncompliance with legal and regulatory requirements must be established.
Has defined policies that have been communicated and that drive behavior	All personnel understand and implement the requirements of the enterprise as defined in policies.
Shows positive behavior toward raising issues or negative outcomes	Whistleblowers are seen as having a positive contribution to the enterprise. The "blame culture" is avoided. Personnel understand the need for risk awareness and reporting possible weaknesses.
Recognizes the value of risk	Personnel understand the importance of the enterprise maintaining awareness and the value that managing risk adds to their role.
Has a transparent and participative culture as an important focus point	Communication is open and overt so that facts are not omitted, misrepresented or understated. The negative impact of hidden agendas is avoided.

Figure 3.35—Relevant Behavior for Risk Governance and Management *(cont.)*

Behavior	Key Objective/Suitable Criteria/Outcome
General (Enterprisewide) Behavior *(cont.)*	
Shows mutual respect	Stakeholders and risk assessors are encouraged to collaborate. People are respected as professionals and treated as experts in their roles.
Accepts ownership of risk	Risk practices are incorporated throughout the enterprise. Accountabilities are cleared and accepted. I&T-related business risk is owned by the business and not viewed solely as the responsibility of the IT department or the risk function.
Allows risk acceptance as a valid option	Management understands the likelihood and consequence of the impact of risk acceptance. The impact is determined to be within the enterprise's risk appetite.
IT Risk Professional-Relevant Behavior	
Effort to understand what risk is for each stakeholder and how it impacts their objectives	Risk professionals understand the commercial reality of the impact of risk. This may include competitive, operational, regulatory and compliance requirements. Although there may be risk common to a certain industry, each enterprise is unique in terms of how these risk items impact specific enterprise objectives.
Awareness and understanding of the risk policy	Alignment among risk capacity, risk appetite and enterprise policy can lead to effective risk strategies.
Collaboration and two-way communication during risk assessment	Risk assessment is fundamentally accurate, complete and addresses stakeholder needs.
Risk appetite that is clear and communicated in a timely fashion with relevant stakeholders	Stakeholders manage risk more effectively and there is appropriate alignment with organizational strategy and objectives.
Policies that reflect risk appetite and risk tolerance	Employees and management operate within risk tolerance. Business lines apply formal risk appetite and tolerance to daily practices. There is a clear process for proposing and making changes to risk appetite levels, with senior management consideration and approval.
Enterprise culture that supports effective risk practice	Stakeholders understand risk from common portfolio views (product, process) and apply risk-based decision making to daily practice.
KRIs used as an early warning	KRIs are associated with valid metrics and can be used as an indicator of process or control failure. KRI metrics are available and accessible for regular reporting and relate to objectives.
Action taken on risk indicators or events that fall outside of appetite and tolerance	Risk indicators are linked to the management risk response and remediation activities.
Management Behavior	
Senior management sets direction and demonstrates visible and genuine support for risk practices.	Quality risk management practices are maintained through genuine support from senior management.
Management engages with all relevant stakeholders to agree on actions and follow up on action plans.	The correct stakeholders are appropriately involved in ensuring timely resolution of issues and the achievement of business plans.
Genuine commitment is obtained and resources are assigned for execution of actions.	Personnel are empowered in executing actions required by risk management decisions.
Management aligns policies and actions to risk appetite.	Management makes appropriate risk decisions complying with policies. Risk-adjusted revenue is in line with management expectations.
Management proactively monitors risk and action plan progress.	Remediation plans are completed within expected business time frames and have a positive impact on enterprise objectives.
Risk trends are reported to management.	The timely reporting of risk trends proactively manages risk and avoids lost opportunities.
Effective risk management is rewarded.	Good risk practice is acknowledged. Employee performance goals and reward structures are set to stimulate effective risk management practices and appropriate execution of mitigation actions.

3.8.2 Influencing Behaviors

This section describes how the behaviors described in the preceding section can be influenced by leadership at different levels of the enterprise: the executive management level, the management of the risk function level and the practitioner level.

Behavior can be influenced through:

- The use of established policies and procedures
- Incentive and reward programs
- Awareness programs

Influencing Behavior Through Communication, Enforcement and Rules

Leadership uses communication, enforcement and rules to influence behaviors in an enterprise. Communication is always essential to influence any kind of behavior. Development and implementation of a risk-aware culture is dependent on the degree of importance of that aspect of a culture. Policies and processes can be used to enforce internal action where regulations are mandated.

Following are some examples of how executive management can use communication, enforcement and rules to influence and promote desirable behavior:

- Management should lead by example as staff takes cues from management for acceptable behavior (tone at the top).
- Risk considerations are embedded in business planning.
- Organizational policies clearly identify the required compliance requirements and ensure that they are enforced.
- Senior management reinforces transparency by cascading appropriate communication throughout the enterprise.
- A zero-tolerance approach for unethical behaviors has clear consequences for misbehavior.
- The CEO communicates to business unit heads to maintain a positive relationship with the assurance function.
- The CEO communicates to business unit heads the need to implement recommendations from risk analysis and propagates the use of root cause analysis.
- Top management approves and communicates a documented security policy, based on which procedures, practices, standards and guidelines are prepared, and shares with all relevant employees.
- Top management encourages active participation in enterprise initiatives.
- Top management establishes a speak-up process to promote and support whistleblowing.
- Regulatory requirements as relevant to ethics are identified and communicated to all employees and are included in the code of ethics.
- A training budget, tools and guidance to support training efforts are provided.
- A communication plan is set up and executed to assist with enterprise change management when a new project, procedure, program, etc., is launched.
- Within an enterprise that has established formal rules, informal structures can be established through positive communication. These informal structures give an extended reach to people who are rarely in contact with each other. This creates a positive synergy in the daily interaction between people.
- The right signals can produce positive effects, while wrong signals can lead to undesirable outcomes. Positive communication can give extraordinary results in working teams.

Influencing Behavior Through Incentives and Rewards

Management influences behavior through measures designed to provide positive reinforcement for desired conduct and negative reinforcement for conduct it wishes to discourage. An absence of rewards inhibits adoption of a risk-aware culture. Business management needs to know that risk-aware behavior will be rewarded; this, in turn, means that executive management must make its intentions clear by encouraging the implementation of appropriate risk responses and promoting attitudes that constitute a culture of risk awareness.

The following incentives and rewards may be used by the various management levels to influence behavior:

- Management performance is aligned to risk management criteria and supports a risk-aware culture.
- Bonuses are paid on executives' conformance to risk practice and expected outcomes; sanctions and fines are imposed when the opposite behavior is demonstrated.
- Engagement with risk is a recognized requirement in evaluation of management performance.
- Management has a policy of recognizing and rewarding employees who obtain specific certification in IT governance, assurance, security and control.
- A structured and visible career progression plan is set up. Doing so lets employees know that if they do the right things or add value, they can climb the ladder of their career.
- Management encourages practitioners among the staff:
  - Active participation is considered as a key attribute in the assessment of individual performance.
  - Ethical behavior is seen a core requirement in performance evaluation.
- Public recognition is a highly valued reward to increase visibility and relationships with managers and peers.
- Financial recognition is commensurate with the level of value:
  - Employees should be part of the whole.
  - Employees should be assured that their role is considered an essential part of the machinery.
  - Personal performance goals, objectives, rewards and sanctions are formalized.

Influencing Behavior Through Raising Awareness

Awareness programs have their place, but they are insufficient by themselves. More than just being aware of I&T information risk, people need to be educated about risk and their role in it. The different management levels in an enterprise can raise awareness through the following means.

Executive management can raise awareness by:

- Visibly supporting the development, execution and completion of training programs to ensure a consistent risk-aware culture
- Regularly communicating to staff the need for risk management throughout the organization to enforce awareness of the risk environment as the basis for decision making

Risk management can raise awareness by:

- Running risk awareness workshops with key stakeholders to communicate key risk impacts and probabilities
- Creating awareness programs that form the basis of what should be done once the risk materializes. These responses (the mitigation or action plans to be taken in the case of risk materializing) must be industry- and organization-specific.

Risk practitioners can raise awareness with:

- Case studies that are used to highlight the business impact of failures in risk awareness or compliance
- Enterprise literature/portal that highlights individuals who champion the desired values

3.9 Component: Services, Infrastructure and Applications

This section identifies and discusses all services, infrastructure and applications that are required to build and sustain effective and efficient risk management in an enterprise. In other words, it lists all services and supporting applications that will support the risk management function.

3.9.1 Services

Services are the support and delivery the I&T risk management function needs to provide, including stakeholder engagement, risk identification, risk analysis, risk monitoring, risk reporting and risk prioritization (**figure 3.36**).

Figure 3.36—Services Provided by the I&T Risk Management Function

Services and Supporting Applications	Description
Program/Project risk advisory services	The services that help to ensure new/changing business strategies, programs, projects, processes and technology maintain an optimized level of risk
I&T incident management services	The services that oversee the response, recovery, monitoring and reporting of incidents
Architecture advisory services	The services that provide subject matter expert guidance to help ensure that the business, data and technology architecture support the risk management objectives of the enterprise
I&T risk intelligence services	The services that provide both tactical and strategic threat, vulnerability and asset intelligence to support enterprise risk decisions
I&T risk management services	The services that provide I&T risk subject matter expertise to support the development and ongoing support of risk management programs within the enterprise
Crisis management services	The people, organizations, processes and technology that help to respond to any type of I&T-related crisis, including those requiring the activation of the business continuity plan (BCP)

3.9.2 Infrastructure

Figure 3.37 identifies and briefly describes some infrastructure components a risk function will need.

Figure 3.37—Infrastructure Components Needed by the Risk Function

Services and Supporting Applications	Description
Data sources	The resources that provide timely risk environment data that support data mining and risk analytics
Infrastructure for knowledge repositories	The resources needed to store intelligence from internal and external sources (e.g., big data)
Intelligence integration architecture	The resources that provide a means of bridging real-time data sources (e.g., systems insight manager [SIM] providers), knowledge repositories (e.g., ERM tools) and risk analytic applications to provide timely decision support

3.9.3 Applications

Figure 3.38 identifies and describes several risk function-related supporting applications.

Figure 3.38—Risk Function-Related Applications	
Services and Supporting Applications	**Description**
Governance, risk and compliance (GRC) tools	A subset of GRC tools, including potential dashboards or balanced scorecards as defined by the enterprise that enable the enterprise to collect, analyze, manage and report risk. These tools aim to communicate the risk in a prioritized order so that the core information can be extracted at a single glance. A risk matrix (risk map) is one such tool, which enables the organization to spot the most critical risk in the repository and their magnitude beyond the approved risk appetite.
Analysis tools	Qualitative and/or quantitative tools to support well-informed risk decision making
Tools for risk communication/reporting	Tools that aim to communicate the findings of risk management
Knowledge repositories	A set of repositories to manage information used to facilitate the risk management analysis and overall process
Business continuity tools	Tools used to help manage risk analysis (BIA, for example) and related activities in a risk management context

Page intentionally left blank

Chapter 4

**COBIT Governance and Management Objectives—
Detailed I&T Risk-Specific Guidance**

4.1 Evaluate, Direct and Monitor (EDM)

01 Ensured Governance Framework Setting and Maintenance

02 Ensured Benefits Delivery

03 Ensured Risk Optimization

05 Ensured Stakeholder Engagement

Page intentionally left blank

Domain: Evaluate, Direct and Monitor Governance Objective: EDM01 — Ensured Governance Framework Setting and Maintenance	Focus Area: I&T Risk
Description	
Analyze and articulate the requirements for the governance of enterprise I&T. Put in place and maintain governance components with clarity of authority and responsibilities to achieve the enterprise's mission, goals and objectives.	
Purpose	
Provide a consistent approach integrated and aligned with the enterprise governance approach. I&T-related decisions are made in line with the enterprise's strategies and objectives and desired value is realized. To that end, ensure that I&T-related processes are overseen effectively and transparently; compliance with legal, contractual and regulatory requirements is confirmed; and the governance requirements for board members are met.	
I&T Risk Focus Area Relevance	
An I&T risk framework is integral and must be aligned with the enterprise risk and I&T governance framework.	

A. Component: Process	
Governance Practice	**Example I&T Risk-specific Metrics**
EDM01.01 Evaluate the governance system. Continually identify and engage with the enterprise's stakeholders, document an understanding of the requirements, and evaluate the current and future design of governance of enterprise I&T.	a. Number of enterprise and IT processes in which I&T risk activities are integrated b. Percent of processes and practices with clear traceability to I&T risk management principles c. Number of stakeholder requirements regarding I&T risk priorities and objectives

I&T Risk-specific Activities (in addition to COBIT 2019 core-model activities)	Capability Level
1. Identify relevant stakeholders with regard to governance of risk (business process owners, management, CRO, etc.).	2
2. Analyze and identify the internal and external environmental risk factors and trends in the business and IT environment that may influence risk governance design.	
3. Determine the significance of I&T risk and its role with respect to the business.	
4. Determine the I&T risk implications as a result of the overall enterprise control environment.	
5. Collect stakeholder requirements with regard to risk priorities and objectives.	
6. Articulate I&T risk decision-making in alignment with the enterprise governance design principles.	3
7. Understand the enterprise's decision-making culture and determine the optimal decision-making model for I&T risk.	
8. Determine the appropriate levels of authority delegation, including threshold rules, for I&T risk decisions.	
9. Coach management decision-makers on the proposed I&T risk analysis approach. Illustrate how risk analysis results can benefit major decisions.	4
10. Describe what level of analysis quality decision makers should expect, how to interpret risk analysis reports, definitions of key terms, and the limitations of measurements and estimates based on incomplete data. Identify gaps with enterprise risk expectations.	

Related Guidance (Standards, Frameworks, Compliance Requirements)	Detailed Reference
CMMI Cybermaturity Platform, 2018	GE.AG Apply Governance System; GE.MG Monitor Governance System
ISO/IEC 38500:2015(E)	5.2 Principle 1: Responsibility (Evaluate)
ITIL V3, 2011	Service Strategy, 2.3 Governance and management systems
National Institute of Standards and Technology Special Publication 800-37, Revision 2 (Draft), May 2018	3.1 Preparation (Tasks 2, 3, 4, 5)

A. Component: Process *(cont.)*	
Governance Practice	**Example I&T Risk-specific Metrics**
EDM01.02 Direct the governance system. Inform leaders on I&T governance principles and obtain their support, buy-in and commitment. Guide the structures, processes and practices for the governance of I&T in line with the agreed governance principles, decision-making models and authority levels. Define the information required for informed decision making.	a. Frequency of I&T risk governance reporting to the executive committee and board

I&T Risk-specific Activities (in addition to COBIT 2019 core-model activities)	Capability Level
1. Communicate I&T risk governance principles and obtain the agreement of the governance body on the way forward to establish informed and committed leadership.	2
2. Establish or delegate the establishment of I&T risk governance structures, processes and practices in line with agreed-upon design principles.	
3. Allocate responsibility, authority and accountability in line with agreed-upon I&T risk governance design principles, decision-making models and delegation. Mandate an enterprisewide risk function.	3
4. Ensure that I&T risk communication and reporting mechanisms provide those responsible for oversight and decision-making with appropriate information in a timely manner.	
5. Communicate the I&T risk management policy, including the principles, organization, roles and responsibilities, communication, and reporting mechanisms.	

Related Guidance (Standards, Frameworks, Compliance Requirements)	Detailed Reference
CMMI Cybermaturity Platform, 2018	GE.DG Direct Governance System
ISF, The Standard of Good Practice for Information Security 2016	SG1.1 Security Governance Framework
ISO/IEC 38500:2015(E)	5.2 Principle 1: Responsibility (Direct)
ISO/IEC 38502:2017(E)	Governance of IT - Framework and model (all chapters)
King IV Report on Corporate Governance for South Africa, 2016	Part 5.4: Governance functional areas - Principle 12
National Institute of Standards and Technology Special Publication 800-53, Revision 5 (Draft), August 2017	3.14 Planning (PL-2, PL-10)

Governance Practice	Example I&T Risk-specific Metrics
EDM01.03 Monitor the governance system. Monitor the effectiveness and performance of the enterprise's governance of I&T. Assess whether the governance system and implemented mechanisms (including structures, principles and processes) are operating effectively and provide appropriate oversight of I&T to enable value creation.	a. Frequency of independent reviews of I&T risk governance documentation b. Number of external/internal audit reports and reviews c. Number of actions needed to rectify deviations d. Number of I&T governance risk issues reported e. Number of adjustments made in reviews to ensure new strategies align to IT governance risk issues f. Number of high I&T risk issues identified through internal/external audits g. Time required to collect and process risk management performance and conformance data

I&T Risk-specific Activities (in addition to COBIT 2019 core-model activities)	Capability Level
1. Periodically assess whether agreed I&T risk governance mechanisms (structures, principles, processes, etc.) are established and operating effectively.	3
2. Identify actions to rectify any deviations found.	
3. Monitor IT and business processes to ensure that these comply with I&T risk activities.	4
4. Document risk governance decision-making through formal meeting minutes and communication of action plans.	

Related Guidance (Standards, Frameworks, Compliance Requirements)	Detailed Reference
ISO/IEC 38500:2015(E)	5.2 Principle 1: Responsibility (Monitor)
National Institute of Standards and Technology Special Publication 800-53, Revision 5 (Draft), August 2017	3.14 Planning (PL-11)

C. Component: Information Flows and Items				
Governance Practice	**I&T Risk-specific Inputs**		**I&T Risk-specific Outputs**	
EDM01.01 Evaluate the governance system.	**From**	**Description**	**Description**	**To**
	Outside COBIT	Internal and external environmental factors	Decision-making model for I&T risk	IRFA EDM01.02 IRFA EDM01.03 IRFA EDM02.04
			Stakeholder requirements with regards to risk priorities and objectives	IRFA EDM01.03 IRFA EDM02.02 IRFA EDM05.02 IRFA MEA01.01
EDM01.02 Direct the governance system.	IRFA EDM01.01	Decision-making model for I&T risk	Risk function mandate	Internal
			IT risk governance principles	
EDM01.03 Monitor the governance system.	IRFA EDM01.01	Decision-making model for I&T risk	Assessment of I&T risk governance mechanisms	Internal
	IRFA EDM01.01	Stakeholder requirements specific to risk priorities and objectives	Formal meeting minutes and communication of action plan	

Related Guidance (Standards, Frameworks, Compliance Requirements)	Detailed Reference
National Institute of Standards and Technology Special Publication 800-37, Revision 2, September 2017	3.1 Preparation (Task 2, 3, 4, 5): Inputs and Outputs

Page intentionally left blank

Domain: Evaluate, Direct and Monitor Governance Objective: EDM02 – Ensured Benefits Delivery	Focus Area: I&T Risk
Description	
Optimize the value to the business from investments in business processes, I&T services and I&T assets.	
Purpose	
Secure optimal value from I&T-enabled initiatives, services and assets; cost-efficient delivery of solutions and services; and a reliable and accurate picture of costs and likely benefits so that business needs are supported effectively and efficiently.	
I&T Risk Focus Area Relevance	
The benefits of the investment in I&T risk management form part of the I&T investment portfolio and are optimized accordingly.	

A. Component: Process	
Governance Practice	**Example I&T Risk-specific Metrics**
EDM02.01 Establish the target investment mix. Review and ensure clarity of the enterprise and I&T strategies and current services. Define an appropriate investment mix based on cost, alignment with strategy, type of benefit for the programs in the portfolio, degree of risk, and financial measures such as cost and expected return on investment (ROI) over the full economic life cycle. Adjust the enterprise and I&T strategies where necessary.	a. Frequency of reviewing the investment mix that attempts to support the right balance among a number of dimensions, including I&T risk initiatives b. Number of changes in strategies within a specific time frame
I&T Risk-specific Activities (in addition to COBIT 2019 core-model activities)	**Capability Level**
1. Identify and record the risk management requirements of stakeholders (such as shareholders, regulators, auditors and customers) for protecting their interests and delivering value.	2
2. Define an investment mix that achieves the right balance among a number of risk dimensions, including I&T risk initiatives.	3
Related Guidance (Standards, Frameworks, Compliance Requirements)	**Detailed Reference**
King IV Report on Corporate Governance for South Africa, 2016	Part 5.5: Stakeholder relationships—Principle 17
The Open Group IT4IT Reference Architecture, Version 2.0	3.2 IT Value Chain and IT4IT Reference Architecture
Governance Practice	**Example I&T Risk-specific Metrics**
EDM02.02 Evaluate value optimization. Continually evaluate the portfolio of I&T-enabled investments, services and assets to determine the likelihood of achieving enterprise objectives and delivering value. Identify and evaluate any changes in direction to management that will optimize value creation.	a. Percent of risk reduction vs. budget deviation (budgeted vs. projected) b. Percent of time KRIs and KPIs have been achieved
I&T Risk-specific Activities (in addition to COBIT 2019 core-model activities)	**Capability Level**
1. Set direction in accordance with the risk management requirements of stakeholders (such as shareholders, regulators, auditors and customers) for protecting their interests and delivering value.	2
2. Evaluate whether the key risk indicators (KRIs) and key performance indicators (KPIs) have been achieved.	3
Related Guidance (Standards, Frameworks, Compliance Requirements)	**Detailed Reference**
COSO Enterprise Risk Management, June 2017	7. Strategy and Objective-Setting—Principle 8
ISF, The Standard of Good Practice for Information Security 2016	SG2.2 Stakeholder Value Delivery
ISO/IEC 38500:2015(E)	5.3 Principle 2: Strategy (Evaluate)
King IV Report on Corporate Governance for South Africa, 2016	Part 5.2: Strategy, performance and reporting—Principle 4
The Open Group IT4IT Reference Architecture, Version 2.0	5. Strategy to Portfolio (S2P) Value Stream

<table>
<tr><th colspan="3">A. Component: Process *(cont.)*</th></tr>
<tr><th>Governance Practice</th><th colspan="2">Example I&T Risk-specific Metrics</th></tr>
<tr><td>**EDM02.03 Direct value optimization.**
Direct value management principles and practices to enable optimal value realization from I&T-enabled investments throughout their full economic life cycle.</td><td colspan="2">a. Amount of investment-related risk exposure, before vs. after mitigation (if a quantitative risk assessment approach exists)</td></tr>
<tr><th colspan="2">I&T Risk-specific Activities (in addition to COBIT 2019 core-model activities)</th><th>Capability Level</th></tr>
<tr><td colspan="2">1. Establish a method of demonstrating the value of risk management (including defining and collecting relevant data) to ensure the efficient use of assets.</td><td>2</td></tr>
<tr><td colspan="2">2. Demonstrate the value of the risk function by highlighting the risk function's contribution to the business objectives.</td><td>3</td></tr>
<tr><th>Related Guidance (Standards, Frameworks, Compliance Requirements)</th><th colspan="2">Detailed Reference</th></tr>
<tr><td>ISO/IEC 38500:2015(E)</td><td colspan="2">5.3 Principle 2: Strategy (Direct)</td></tr>
<tr><th>Governance Practice</th><th colspan="2">Example I&T Risk-specific Metrics</th></tr>
<tr><td>**EDM02.04 Monitor value optimization.**
Monitor key goals and metrics to determine whether the enterprise receives expected value and benefit from I&T-enabled investments and services. Identify significant issues and consider corrective actions.</td><td colspan="2">a. Level of stakeholder satisfaction with the I&T risk management measures in place, based on surveys</td></tr>
<tr><th colspan="2">I&T Risk-specific Activities (in addition to COBIT 2019 core-model activities)</th><th>Capability Level</th></tr>
<tr><td colspan="2">1. Establish KRIs.</td><td>2</td></tr>
<tr><td colspan="2">2. Monitor the value of risk initiatives and compare the results to stakeholder requirements that were set to ensure value delivery.</td><td rowspan="2">4</td></tr>
<tr><td colspan="2">3. Use business-focused methods of reporting on the added value of risk management initiatives.</td></tr>
<tr><td colspan="2">4. Upon review of the KRIs, ensure that appropriate management corrective action is initiated and controlled.</td><td rowspan="2">5</td></tr>
<tr><td colspan="2">5. Upon review of the KRIs, take appropriate management action as required to ensure that value is optimized.</td></tr>
<tr><th>Related Guidance (Standards, Frameworks, Compliance Requirements)</th><th colspan="2">Detailed Reference</th></tr>
<tr><td>ISO/IEC 38500:2015(E)</td><td colspan="2">5.3 Principle 2: Strategy (Monitor)</td></tr>
</table>

C. Component: Information Flows and Items				
Governance Practice	**I&T Risk-specific Inputs**		**I&T Risk-specific Outputs**	
	From	**Description**	**Description**	**To**
EDM02.01 Establish the target investment mix.	There are no I&T risk-specific inputs or outputs for this practice.			
EDM02.02 Evaluate value optimization.	IRFA EDM01.01	Stakeholder requirements specific to risk priorities and objectives	Formal documentation of stakeholder requirements and direction with regard to risk tolerance levels in I&T investment policy	IRFA EDM02.03 IRFA EDM02.04
EDM02.03 Direct value optimization.	IRFA EDM01.02	Formal documentation of stakeholder requirements and direction with regard to risk tolerance levels in I&T investment policy	Formal documentation of risk function's contribution to business objectives	Internal
EDM02.04 Monitor value optimization.	IRFA EDM01.01	Decision-making model for I&T risk	Risk function mandate	Internal
	IRFA EDM02.02	Formal documentation of stakeholder requirements and direction with regard to risk tolerance levels in I&T investment policy	IT risk governance principles	

Related Guidance (Standards, Frameworks, Compliance Requirements)	Detailed Reference
No related guidance for this component	

Page intentionally left blank

Domain: Evaluate, Direct and Monitor Governance Objective: EDM03 – Ensured Risk Optimization	Focus Area: I&T Risk
Description	
Ensure that the enterprise's risk appetite and tolerance are understood, articulated and communicated, and that risk to enterprise value related to the use of I&T is identified and managed.	
Purpose	
Ensure that I&T-related enterprise risk does not exceed the enterprise's risk appetite and risk tolerance, the impact of I&T risk to enterprise value is identified and managed, and the potential for compliance failures is minimized.	
I&T Risk Focus Area Relevance	
Ensure business risk associated with the use, ownership, operation, involvement, influence and adoption of I&T within an enterprise to minimize the impact of I&T-related events that could potentially impact the business.	

A. Component: Process		
Governance Practice	**Example I&T Risk-specific Metrics**	
EDM03.01 Evaluate risk management. Continually examine and evaluate the effect of risk on the current and future use of I&T in the enterprise. Consider whether the enterprise's risk appetite is appropriate and ensure that risk to enterprise value related to the use of I&T is identified and managed.	a. Number of revisited I&T risk issues when the issue exceeds the risk appetite	
I&T Risk-specific Activities (in addition to COBIT 2019 core-model activities)		**Capability Level**
Additional I&T risk-specific activities have not been identified for this practice. The COBIT 2019 core activities are applicable.		N/A
Related Guidance (Standards, Frameworks, Compliance Requirements)	**Detailed Reference**	
COSO Enterprise Risk Management, June 2017	Strategy and Objective-Setting—Principles 6 and 7; 9. Review and Revision—Principle 16	
Governance Practice	**Example I&T Risk-specific Metrics**	
EDM03.02 Direct risk management. Direct the establishment of risk management practices to provide reasonable assurance that I&T risk management practices are appropriate, and that actual I&T risk does not exceed the board's risk appetite.	Additional I&T risk-specific metrics have not been identified for this practice. The COBIT 2019 core metrics are applicable.	
I&T Risk-specific Activities (in addition to COBIT 2019 core-model activities)		**Capability Level**
Additional I&T risk-specific activities have not been identified for this practice. The COBIT 2019 core activities are applicable.		N/A
Related Guidance (Standards, Frameworks, Compliance Requirements)	**Detailed Reference**	
CMMI Cybermaturity Platform, 2018	RS.AS Apply Risk Management Strategy; BC.RO Determine Strategic Risk Objectives	
ISF, The Standard of Good Practice for Information Security 2016	IR1.1 Information Risk Assessment—Management Approach	
King IV Report on Corporate Governance for South Africa, 2016	Part 5.4: Governance functional areas—Principle 11	
National Institute of Standards and Technology Special Publication 800-37, Revision 2 (Draft), May 2018	3.5 Assessment (Task 2)	

A. Component: Process *(cont.)*		
Governance Practice	**Example I&T Risk-specific Metrics**	
EDM03.03 Monitor risk management. Monitor the key goals and metrics of the risk management processes. Determine how deviations or problems will be identified, tracked and reported for remediation.	Additional I&T risk-specific metrics have not been identified for this practice. The COBIT 2019 core metrics are applicable.	
I&T Risk-specific Activities (in addition to COBIT 2019 core-model activities)		**Capability Level**
Additional I&T risk-specific activities have not been identified for this practice. The COBIT 2019 core activities are applicable.		N/A
Related Guidance (Standards, Frameworks, Compliance Requirements)	**Detailed Reference**	
COSO Enterprise Risk Management, June 2017	9. Review and Revision—Principle 17	
National Institute of Standards and Technology Special Publication 800-37, Revision 2 (Draft), May 2018	3.1 Preparation (Task 7); 3.5 Assessment (Task 1); 3.6 Authorization (Task 1)	
The Open Group IT4IT Reference Architecture, Version 2.0	6. Requirement to Deploy (R2D) Value Stream; 7. Request to Fulfill (R2F) Value Stream	

C. Component: Information Flows and Items				
Governance Practice	**I&T Risk-specific Inputs**		**I&T Risk-specific Outputs**	
	From	**Description**	**Description**	**To**
EDM03.01 Evaluate risk management.	Outside COBIT	Threat intel	Risk universe, appetite and tolerance	All
			Key risk indicators	IRFA APO01.01 IRFA BAI03.10
EDM03.02 Direct risk management.	There are no I&T risk-specific inputs for this practice.		Risk taxonomy	All
			Risk communication plan	IRFA APO01.02
EDM03.03 Monitor risk management.	IRFA APO07.03	Risk awareness program	There are no I&T risk-specific outputs for this practice.	
Related Guidance (Standards, Frameworks, Compliance Requirements)	**Detailed Reference**			
National Institute of Standards and Technology Special Publication 800-37, Revision 2, September 2017	3.1 Preparation (Task 7): Inputs and Outputs; 3.5 Assessment (Tasks 1, 2): Inputs 2, and Outputs; 3.6 Authorization (Task 1): Inputs and Outputs			

Domain: Evaluate, Direct and Monitor Governance Objective: EDM05 — Ensured Stakeholder Engagement	Focus Area: I&T Risk
Description	
Ensure that stakeholders are identified and engaged in the I&T governance system and that enterprise I&T performance and conformance measurement and reporting are transparent, with stakeholders approving the goals and metrics and necessary remedial actions.	
Purpose	
Ensure that stakeholders are supportive of the I&T strategy and road map, communication to stakeholders is effective and timely, and the basis for reporting is established to increase performance. Identify areas for improvement, and confirm that I&T-related objectives and strategies are in line with the enterprise's strategy.	
I&T Risk Focus Area Relevance	
Stakeholders need to be engaged for I&T risk-related performance and conformance matters.	

A. Component: Process		
Governance Practice	**Example I&T Risk-specific Metrics**	
EDM05.01 Evaluate stakeholder engagement and reporting requirements. Continually examine and evaluate current and future requirements for stakeholder engagement and reporting (including reporting mandated by regulatory requirements), and communication to other stakeholders. Establish principles for engaging and communicating with stakeholders.	a. Number of stakeholders that currently receive I&T risk insights b. Percent of stakeholders covered in I&T risk reporting requirements c. Number of external and internal stakeholders that currently receive I&T risk insights	
I&T Risk-specific Activities (in addition to COBIT 2019 core-model activities)		**Capability Level**
1. Determine the audience, including internal and external individuals or groups, for communication and reporting on I&T risk-related activities.		2
2. Examine and make a judgment on the current and future reporting requirements relating to I&T risk within the enterprise (regulation, legislation, common law, contractual), including extent and frequency.		
3. Identify the means and channels to communicate risk issues.		3
Related Guidance (Standards, Frameworks, Compliance Requirements)	**Detailed Reference**	
CMMI Cybermaturity Platform, 2018	SR.DR Direct Stakeholder Communication and Reporting	
Governance Practice	**Example I&T Risk-specific Metrics**	
EDM05.02 Direct stakeholder engagement, communication and reporting. Ensure the establishment of effective stakeholder involvement, communication and reporting, including mechanisms for ensuring the quality and completeness of information, overseeing mandatory reporting, and creating a communication strategy for stakeholders.	a. Percent of stakeholder satisfaction with I&T risk management communication and reporting b. Number of stakeholders involved in I&T risk management	
I&T Risk-specific Activities (in addition to COBIT 2019 core-model activities)		**Capability Level**
1. Prioritize reporting on risk issues to stakeholders.		2
2. Produce regular risk and risk management status reports for stakeholders.		3
3. Ensure status reports include risk management activities, risk treatment against target dates, performance, achievements, risk profile, business benefits, 'hot topics' (e.g., cloud, consumer products), outstanding risk and capability gaps.		
4. Evaluate and, where appropriate, incorporate stakeholder feedback on risk management reports.		
Related Guidance (Standards, Frameworks, Compliance Requirements)	**Detailed Reference**	
CMMI Cybermaturity Platform, 2018	SR.AR Apply Stakeholder Reporting Requirements	
King IV Report on Corporate Governance for South Africa, 2016	Part 5.5: Stakeholder relationships—Principle 16	
King IV Report on Corporate Governance for South Africa, 2016	Part 5.2: Strategy, performance and reporting—Principle 5	
National Institute of Standards and Technology Framework for Improving Critical Infrastructure Cybersecurity V1.1, April 2018	3.3 Communicating Cybersecurity Requirements with Stakeholders	

A. Component: Process *(cont.)*		
Governance Practice	**Example I&T Risk-specific Metrics**	
EDM05.03 Monitor stakeholder engagement. Monitor stakeholder engagement levels and the effectiveness of stakeholder communication. Assess mechanisms for ensuring accuracy, reliability and effectiveness, and ascertain whether the requirements of different stakeholders in terms of reporting and communication are met.	a. Percent of I&T risk management reports containing inaccuracies b. Percent of I&T risk management reports delivered on time	
I&T Risk-specific Activities (in addition to COBIT 2019 core-model activities)		**Capability Level**
1. Establish risk monitoring and reporting using, for example, KRIs and KPIs for risk and risk management that are based on metrics and measurements in the MEA domain.		4
Related Guidance (Standards, Frameworks, Compliance Requirements)	**Detailed Reference**	
CMMI Cybermaturity Platform, 2018	SR.MC Monitor Stakeholder Communication	

C. Component: Information Flows and Items				
Governance Practice	**I&T Risk-specific Inputs**		**I&T Risk-specific Outputs**	
	From	**Description**	**Description**	**To**
EDM05.01 Evaluate stakeholder engagement and reporting requirements.	Outside COBIT	Evaluation of enterprise reporting requirements	Evaluation of risk reporting requirements and communication channels	IRFA EDM05.03
EDM05.02 Direct stakeholder engagement, communication and reporting.	IRFA EDM01.01	Stakeholder requirements specific to risk priorities and objectives	Summary of activities to the enterprise risk committee (ERC)	Internal
EDM05.03 Monitor stakeholder engagement.	IRFA EDM05.01	Evaluation of risk reporting requirements and communication channels	Risk monitoring and reporting	IRFA APO01.02 IRFA APO01.11
Related Guidance (Standards, Frameworks, Compliance Requirements)	**Detailed Reference**			
No related guidance for this component				

4.2 Align, Plan and Organize (APO)

Page intentionally left blank

Domain: Align, Plan and Organize Management Objective: APO01 — Managed I&T Management Framework	Focus Area: I&T Risk
Description	
Design the management system for enterprise I&T based on enterprise goals and other design factors. Based on this design, implement all required components of the management system.	
Purpose	
Implement a consistent management approach for enterprise governance requirements to be met, covering governance components such as management processes; organizational structures; roles and responsibilities; reliable and repeatable activities; information flows and items; policies and procedures; people, skills and competencies; culture, ethics and behavior; and competencies and items; policies and procedures; information flows and items; people, skills and competencies, culture, ethics and behavior and services, infrastructure and applications.	
I&T Risk Focus Area Relevance	
I&T risk management is an integral part of the enterprise's overall I&T management framework.	

A. Component: Process	
Management Practice	**Example I&T Risk-specific Metrics**
APO01.01 Design the management system for enterprise I&T. Design a management system tailored to the needs of the enterprise. Management needs of the enterprise are defined through the use of the goals cascade and by application of design factors. Ensure the governance components are integrated and aligned with the enterprise's governance and management philosophy and operating style.	a. Percent of key risk indicators (KRIs) linked to governance components b. Percent of governance activities reflected in I&T risk monitoring and reporting c. Percent of governance activities monitored through KRIs d. Percent of I&T risk processes documented in the enterprise

I&T Risk-specific Activities (in addition to COBIT 2019 core-model activities)	Capability Level
1. Ensure that roles and responsibilities have been defined and are current.	2
2. Ensure that governance/risk committee meetings are held periodically (e.g., weekly, monthly and annually).	
3. Ensure that the information communicated encompasses I&T risk management practices and activities.	3
4. Ensure that governance components are taking into account the I&T risk management objectives and are aligned with the organization's risk tolerance.	
5. Ensure that resources are certified in functions where they may be required (i.e., information risk, enterprise governance or information security).	
6. Ensure that management reports reflect the actions taken as a result of review by executive management and the board of directors.	

Related Guidance (Standards, Frameworks, Compliance Requirements)	Detailed Reference
COSO Enterprise Risk Management, June 2017	7. Strategy and Objective-Setting—Principle 9
ISO/IEC 27001:2013/Cor.2:2015(E)	International standard for establishing, implementing and maintaining a management system (all chapters)
ITIL V3, 2011	Service Strategy, 2.3 Governance and management systems

A. Component: Process *(cont.)*		
Management Practice	**Example I&T Risk-specific Metrics**	
APO01.02 Communicate management objectives, direction and decisions made. Communicate awareness and promote understanding of alignment and I&T objectives to stakeholders throughout the enterprise. Communicate at regular intervals on important I&T-related decisions and their impact for the organization.	a. Percent of communication taking into account the I&T risk tolerances and I&T risk management objectives b. Number of I&T risk-related communications	
I&T Risk-specific Activities (in addition to COBIT 2019 core-model activities)		**Capability Level**
1. Communicate I&T risk management objectives and I&T risk tolerances.		3
2. Communicate the impact of management decisions on I&T risk levels.		
3. Communicate current risk levels against I&T risk tolerances.		
4. Communicate the status of I&T risk remediation actions.		
Related Guidance (Standards, Frameworks, Compliance Requirements)	**Detailed Reference**	
No related guidance for this management practice		
Management Practice	**Example I&T Risk-specific Metrics**	
APO01.03 Implement management processes (to support the achievement of governance and management objectives). Define target process capability levels and implementation priority based on the management system design.	a. Percent of target governance process model components aiming to improve I&T risk management	
I&T Risk-specific Activities (in addition to COBIT 2019 core-model activities)		**Capability Level**
1. Analyze the gap between the target and current I&T risk management practices and activities.		2
2. Ensure that the target governance process model takes into account the I&T risk management objectives.		3
3. Ensure that I&T risk levels are an input for prioritization of the target governance process model components.		
4. Include I&T risk management practices and activities into the road map.		4
Related Guidance (Standards, Frameworks, Compliance Requirements)	**Detailed Reference**	
No related guidance for this management practice		
Management Practice	**Example I&T Risk-specific Metrics**	
APO01.04 Define and implement the organizational structures. Put in place the required internal and extended organizational structures (e.g., committees) per the management system design, enabling effective and efficient decision making. Ensure that required technology and information knowledge is included in the composition of management structures.	a. Percent of I&T risk function representation in business committees b. Percent of organizational structures with defined I&T risk management responsibilities c. Number of governance bodies with an I&T risk management agenda	
I&T Risk-specific Activities (in addition to COBIT 2019 core-model activities)		**Capability Level**
1. Ensure a proper mix of expertise within the organizational structure.		2
2. Establish escalation mechanisms to enable reporting of risk to the appropriate levels of the enterprise.		3
3. Ensure that I&T risk-related information is available to relevant organizational structures.		
Related Guidance (Standards, Frameworks, Compliance Requirements)	**Detailed Reference**	
No related guidance for this management practice		

A. Component: Process *(cont.)*		
Management Practice	**Example I&T Risk-specific Metrics**	
AP001.05 Establish roles and responsibilities. Define and communicate roles and responsibilities for enterprise I&T, including authority levels, responsibilities and accountability.	a. Percent of roles with I&T risk management responsibilities defined	
I&T Risk-specific Activities (in addition to COBIT 2019 core-model activities)		**Capability Level**
1. Define I&T risk management-related responsibilities and accountabilities for each role.		2
2. Determine if a communication plan exists.		
3. Determine if roles and responsibilities are communicated.		
Related Guidance (Standards, Frameworks, Compliance Requirements)	**Detailed Reference**	
No related guidance for this management practice		
Management Practice	**Example I&T Risk-specific Metrics**	
AP001.06 Optimize the placement of the IT function. Position the IT capabilities in the overall organizational structure to reflect the strategic importance and operational dependency of IT within the enterprise. The reporting line of the CIO and representation of IT within senior management should be commensurate with the importance of I&T within the enterprise.	a. Time required for I&T to escalate key I&T risk to senior management b. Degree of organizational distance between I&T and risk decision makers	
I&T Risk-specific Activities (in addition to COBIT 2019 core-model activities)		**Capability Level**
1. Ensure that placement of the IT function allows for timely and accurate decisions for I&T risk management, including I&T risk tolerance, I&T risk acceptance, and escalation of I&T risks.		3
Related Guidance (Standards, Frameworks, Compliance Requirements)	**Detailed Reference**	
ISO/IEC 27002:2013/Cor.2:2015(E)	8.2 Information classification	
Management Practice	**Example I&T Risk-specific Metrics**	
AP001.07 Define information (data) and system ownership. Define and maintain responsibilities for ownership of information (data) and information systems. Ensure that owners classify information and systems and protect them in line with their classification.	a. Percent of information and system owners formally taking into account the associated I&T risk b. Percent of information and system owners trained for I&T risk assessment and mitigation c. Percent of assigned classifications aligned with other sources of I&T risk assessments (e.g., BIA) d. Number of classification types	
I&T Risk-specific Activities (in addition to COBIT 2019 core-model activities)		**Capability Level**
1. Train information and system owners to support I&T risk assessment and classification.		2
2. Ensure that information and system classifications are aligned with other risk assessments.		3
Related Guidance (Standards, Frameworks, Compliance Requirements)	**Detailed Reference**	
No related guidance for this management practice		
Management Practice	**Example I&T Risk-specific Metrics**	
AP001.08 Define target skills and competencies. Define the required skills and competencies to achieve relevant management objectives.	a. Percent of I&T risk management roles with confirmed I&T risk management skills and competencies b. Frequency of skills gap analysis	
I&T Risk-specific Activities (in addition to COBIT 2019 core-model activities)		**Capability Level**
1. Ensure that sufficient skills and competencies exist to achieve I&T risk management objectives.		2
Related Guidance (Standards, Frameworks, Compliance Requirements)	**Detailed Reference**	
No related guidance for this management practice		

A. Component: Process *(cont.)*		
Management Practice	**Example I&T Risk-specific Metrics**	
APO01.09 Define and communicate policies and procedures. Put in place procedures to maintain compliance with and performance measurement of policies and other components of the control framework. Enforce the consequences of noncompliance or inadequate performance. Track trends and performance and consider these in the future design and improvement of the control framework.	a. Number of employees who have attended I&T risk management training b. Number of employees who have completed annual attestation to reading the policies and procedures c. Percent of policies with defined I&T risk management objectives d. Percent of team evaluations negatively affected by identified noncompliance issues	
I&T Risk-specific Activities (in addition to COBIT 2019 core-model activities)		**Capability Level**
1. Define the risk management objectives of policies and procedures.		2
2. Ensure that policies drive I&T risk management expectations.		3
Related Guidance (Standards, Frameworks, Compliance Requirements)	**Detailed Reference**	
No related guidance for this management practice		
Management Practice	**Example I&T Risk-specific Metrics**	
APO01.10 Define and implement infrastructure, services and applications to support the governance and management system. Define and implement infrastructure, services and applications to support the governance and management system (e.g., architecture repositories, risk management system, project management tools, cost-tracking tools and incident monitoring tools).	a. Number of critical systems, infrastructure tools and applications that have undergone an I&T risk assessment b. Number of critical systems, infrastructure tools and applications without I&T risk treatment c. Number of critical systems requiring I&T risk acceptance by the business d. Number of nonintegrated governance, risk and control (GRC) solutions/applications in place	
I&T Risk-specific Activities (in addition to COBIT 2019 core-model activities)		**Capability Level**
1. Identify I&T risk issues resulting from chosen infrastructure, services and applications.		2
2. Define and implement integrated GRC solutions capable of creating efficiency and collaboration between the risk management function and other lines of defense (audit, process transformation, internal control).		
Related Guidance (Standards, Frameworks, Compliance Requirements)	**Detailed Reference**	
No related guidance for this management practice		
A. Component: Process *(cont.)*		
Management Practice	**Example I&T Risk-specific Metrics**	
APO01.11 Manage continual improvement of the I&T management system. Continually improve processes and other management system components to ensure that they can deliver against governance and management objectives. Consider COBIT implementation guidance, emerging standards, compliance requirements, automation opportunities and the feedback of stakeholders.	a. Date of last policy/procedure review and approval b. Number of I&T risk metrics c. Number of I&T risk-mitigating actions triggered by risk monitoring	
I&T Risk-specific Activities (in addition to COBIT 2019 core-model activities)		**Capability Level**
1. Regularly assess I&T risk mitigation performance.		3
2. Assess the quality of risk management processes.		
3. Identify options for improving risk management based on risk mitigation performance.		4
4. Validate risk measurement data and calculations.		
Related Guidance (Standards, Frameworks, Compliance Requirements)	**Detailed Reference**	
ITIL V3, 2011	Continual Service Improvement, 4.1 The 7-Step Improvement Process	

C. Component: Information Flows and Items				
Management Practice	**I&T Risk-specific Inputs**		**I&T Risk-specific Outputs**	
	From	**Description**	**Description**	**To**
APO01.01 Design the management system for enterprise I&T.	IRFA EDM03.01	Key risk indicators	Data, process and control owners	IRFA APO01.07 IRFA APO14.01
	IRFA APO07.01	Risk function requirements for the staffing process	I&T risk appetite and tolerances	IRFA APO01.03
	IRFA APO01.03	Updated management system based on risk appetite	Risk quantification approach (e.g., quantify risk impact)	IRFA APO06.01
APO01.02 Communicate management objectives, direction and decisions made.	IRFA EDM05.03	Risk monitoring and reporting	There are no I&T risk-specific outputs for this practice.	
	IRFA EDM03.02	Risk communication plan		
APO01.03 Implement management processes (to support the achievement of governance and management objectives).	IRFA APO01.01	I&T risk objectives and tolerances	Updated management system based on risk appetite	IRFA APO01.01 IRFA APO01.11
APO01.04 Define and implement the organizational structures.	There are no I&T risk-specific inputs for this practice.		A defined risk function	Internal
APO01.05 Establish roles and responsibilities.	There are no I&T risk-specific inputs for this practice.		Established risk roles and responsibility descriptions	Internal
APO01.06 Optimize the placement of the IT function.	Outside COBIT	Enterprise risk strategy	There are no I&T risk-specific outputs for this practice.	
APO01.07 Define information (data) and system ownership.	IRFA APO01.01	Data, process and control owners	There are no I&T risk-specific outputs for this practice.	
APO01.08 Define target skills and competencies.	There are no I&T risk-specific inputs for this practice.		Risk related skills requirements specifications	Internal
APO01.09 Define and communicate policies and procedures.	IRFA MEA03.01	External risk management compliance requirements	There are no I&T risk-specific outputs for this practice.	
APO01.10 Define and implement infrastructure, services and applications to support the governance and management system.	There are no I&T risk-specific inputs or outputs for this practice.			
APO01.11 Manage continual improvement of the I&T management system.	IRFA APO01.03	Updated management system based on risk appetite	There are no I&T risk-specific outputs for this practice.	
	IRFA EDM05.03	Risk monitoring and reporting		
Related Guidance (Standards, Frameworks, Compliance Requirements)		**Detailed Reference**		
No related guidance for this management practice				

Page intentionally left blank

Domain: Align, Plan and Organize Management Objective: APO02 – Managed Strategy	Focus Area: I&T Risk
Description	
Provide a holistic view of the current business and I&T environment, the future direction, and the initiatives required to migrate to the desired future environment. Ensure that the desired level of digitization is integral to the future direction and the I&T strategy. Assess the organization's current digital maturity and develop a road map to close the gaps. With the business, rethink internal operations as well as customer-facing activities. Ensure focus on the transformation journey across the organization. Leverage enterprise architecture building blocks, governance components and the organization's ecosystem, including externally provided services and related capabilities, to enable reliable but agile and efficient response to strategic objectives.	
Purpose	
Support the digital transformation strategy of the organization and deliver the desired value through a road map of incremental changes. Use a holistic I&T approach, ensuring that each initiative is clearly connected to an overarching strategy. Enable change in all different aspects of the organization, from channels and processes to data, culture, skills, operating model and incentives.	
I&T Risk Focus Area Relevance	
An I&T risk management strategy must be defined, aligned and integrated with the enterprise risk management strategy while taking the I&T strategy into consideration.	

A. Component: Process		
Management Practice	**Example I&T Risk-specific Metrics**	
APO02.01 Understand enterprise context and direction. Understand the enterprise context (industry drivers, relevant regulations, basis for competition), its current way of working and its ambition level in terms of digitization.	a. Stakeholder approval of the I&T risk function charter b. Number of I&T risk policies approved/reviewed by the governance body c. Percent of IT projects that have I&T risk requirements championed by business owners	
I&T Risk-specific Activities (in addition to COBIT 2019 core-model activities)		**Capability Level**
1. Understand how the I&T risk function should support the overall enterprise objectives and protect stakeholder interests by taking into account the need to manage I&T risk while meeting regulatory requirements and adding value to the enterprise.		2
2. Understand the current enterprise architecture and identify potential I&T risk function coverage gaps.		
3. Understand inherent risk in the environment (e.g., industry, geography) in which the enterprise is operating.		
Related Guidance (Standards, Frameworks, Compliance Requirements)	**Detailed Reference**	
COSO Enterprise Risk Management, June 2017	7. Strategy and Objective-Setting—Principle 6	
Management Practice	**Example I&T Risk-specific Metrics**	
APO02.02 Assess current capabilities, performance and digital maturity of the enterprise. Assess the performance of current I&T services and develop an understanding of current business and I&T capabilities (both internal and external). Assess current digital maturity of the enterprise and its appetite for change.	a. Percent of stakeholders satisfied with current I&T risk capabilities b. Percent and number of initiatives for which a risk metric (e.g., risk impact, KRI) has been calculated c. Percent of initiatives that achieved the expected return on investment (ROI)	
I&T Risk-specific Activities (in addition to COBIT 2019 core-model activities)		**Capability Level**
1. Develop a baseline of the current I&T risk management practices and I&T risk environment against which future requirements can be compared.		2
2. Ensure that I&T risk issues arising from current and declining technologies are identified in the I&T risk management process.		3
3. Identify gaps between current business and I&T capabilities and services and reference standards and best practices, competitor business and I&T capabilities, and comparative benchmarks of best practice and emerging I&T service provision.		
4. Identify issues, strengths, opportunities and threats in the current environment, capabilities and services to understand current performance, and identify areas for improvement in terms of I&T's contribution to enterprise objectives.		4
Related Guidance (Standards, Frameworks, Compliance Requirements)	**Detailed Reference**	
COSO Enterprise Risk Management, June 2017	7. Strategy and Objective-Setting—Principle 6; 9. Review and Revision—Principle 15	

A. Component: Process *(cont.)*		
Management Practice	**Example I&T Risk-specific Metrics**	
APO02.03 Define target digital capabilities. Based on the understanding of enterprise context and direction, define the target I&T products and services and required capabilities. Consider reference standards, best practices and validated emerging technologies.	a. Number of threats from declining, current and newly acquired technology b. Percent of I&T risk identified before development of capabilities	
I&T Risk-specific Activities (in addition to COBIT 2019 core-model activities)		**Capability Level**
1. Ensure proper I&T risk analysis is done when defining target I&T capabilities.		2
2. Agree on the impact of I&T risk on changes of the enterprise architecture (business, information, data, applications and technology domains) and business and I&T processes and procedures.		3
3. Agree on the impact of risk on changes of the IT organization structure, I&T service providers, governance of I&T, and I&T skills and competencies.		
4. Ensure that risk issues arising from potential and newly acquired technologies are identified in the risk management process.		
Related Guidance (Standards, Frameworks, Compliance Requirements)	**Detailed Reference**	
No related guidance for this management practice		
Management Practice	**Example I&T Risk-specific Metrics**	
APO02.04 Conduct a gap analysis. Identify gaps between current and target environments and describe the high-level changes in the enterprise architecture.	a. Number of accepted I&T risk issues b. Number of significant gaps between compliance requirements and the current environment c. Number of I&T risk exceptions d. Number of times I&T risk exceptions were renewed	
I&T Risk-specific Activities (in addition to COBIT 2019 core-model activities)		**Capability Level**
1. Identify all gaps to be closed and changes required to realize the target environment, in light of the I&T risk management processes, requirements and I&T risk appetite of the enterprise.		3
2. Examine the current environment with respect to regulations and compliance requirements.		
3. Ensure that the IT risk acceptance process is followed when a decision has been made to accept an I&T risk.		4
Related Guidance (Standards, Frameworks, Compliance Requirements)	**Detailed Reference**	
No related guidance for this management practice		
Management Practice	**Example I&T Risk-specific Metrics**	
APO02.05 Define the strategic plan and road map. Develop a holistic digital strategy, in cooperation with relevant stakeholders, and detail a road map that defines the incremental steps required to achieve the goals and objectives. Ensure focus on the transformation journey through the appointment of a person who helps spearhead the digital transformation and drives alignment between business and I&T.	a. Number of I&T risk requirements in the I&T strategic plan b. Percent of risk information available to support implementation of the strategic plan c. Percent of strategic plan and road map items with calculated I&T risk impact d. Accuracy of the estimation of the I&T risk mitigation after execution of the strategic plan and the road map	
I&T Risk-specific Activities (in addition to COBIT 2019 core-model activities)		**Capability Level**
1. Define the I&T risk strategy and align it with business strategies and the enterprise's overall objectives and I&T risk appetite.		3
2. Ensure that the current I&T strategic plan and road map take into account the I&T risk requirements.		
Related Guidance (Standards, Frameworks, Compliance Requirements)	**Detailed Reference**	
ISF, The Standard of Good Practice for Information Security 2016	SG2.1 Information Security Strategy	
ITIL V3, 2011	Service Strategy, 4.1 Strategy management for IT services	

A. Component: Process *(cont.)*	
Management Practice	**Example I&T Risk-specific Metrics**
APO02.06 Communicate the I&T strategy and direction. Create awareness and understanding of the business and I&T objectives and direction, as captured in the I&T strategy, through communication to appropriate stakeholders and users throughout the enterprise.	a. Enterprise stakeholder satisfaction survey feedback on the effectiveness of the I&T risk management strategy b. Percent of stakeholders aware of the I&T risk management strategy c. Percent of strategic objectives and road map items with estimated I&T risk impact or I&T risk mitigation targets

I&T Risk-specific Activities (in addition to COBIT 2019 core-model activities)	Capability Level
1. Develop the I&T risk management plan based on the I&T risk management strategy.	3
2. Gain approval from key stakeholders on the I&T risk management strategy.	
3. Communicate the risk management strategy and plan to all relevant stakeholders.	

Related Guidance (Standards, Frameworks, Compliance Requirements)	Detailed Reference
No related guidance for this management practice	

C. Component: Information Flows and Items				
Management Practice	**I&T Risk-specific Inputs**		**I&T Risk-specific Outputs**	
	From	**Description**	**Description**	**To**
APO02.01 Understand enterprise context and direction.	There are no I&T risk-specific inputs for this practice.		Listing of potential risk function coverage gaps	IRFA APO02.02
APO02.02 Assess current capabilities, performance and digital maturity of the enterprise.	IRFA APO02.01	Listing of potential risk function coverage gaps	Risk function capabilities	IRFA APO02.03
			Baseline of the current business and IT environment	IRFA APO02.04
APO02.03 Define target digital capabilities.	IRFA APO02.02	Risk function capabilities	Risk management requirements in target IT capabilities	IRFA APO02.04 IRFA APO02.05
APO02.04 Conduct a gap analysis.	IRFA APO02.02	Baseline of the current business and IT environment	Risk management coverage gaps to be closed	IRFA APO02.05
	IRFA APO02.03	Risk management requirements in target IT capabilities		
	IRFA APO05.01	Business case for I&T Risk initiatives		
	Outside COBIT	Regulation and compliance requirements for enterprise I&T		
APO02.05 Define the strategic plan and road map.	IRFA APO02.03	Risk management requirements in target IT capabilities	Updated I&T strategic plan and road map taking into account the risk management requirements	Internal
	IRFA APO02.04	Risk management coverage gaps to be closed	Risk management strategy	IRFA APO02.06
APO02.06 Communicate the I&T strategy and direction.	IRFA APO02.05	Risk management strategy	Risk management plan	IRFA APO07.01 IRFA APO11.01 IRFA DSS04.02

Related Guidance (Standards, Frameworks, Compliance Requirements)	Detailed Reference
ITIL V3, 2011	Service strategy, 3.9 Service strategy inputs and outputs

Page intentionally left blank

Domain: Align, Plan and Organize Management Objective: APO05 — Managed Portfolio	Focus Area: I&T Risk
Description	
Execute the strategic direction set for investments in line with the enterprise architecture vision and I&T road map. Consider the different categories of investments and the resources and funding constraints. Evaluate, prioritize and balance programs and services, managing demand within resource and funding constraints, based on their alignment with strategic objectives, enterprise worth and risk. Move selected programs into the active products or services portfolio for execution. Monitor the performance of the overall portfolio of products and services and programs, proposing adjustments as necessary in response to program, product or service performance or changing enterprise priorities.	
Purpose	
Optimize the performance of the overall portfolio of programs in response to individual program, product and service performance and changing enterprise priorities and demand.	
I&T Risk Focus Area Relevance	
I&T risk management programs and services are integrated into and form part of the enterprise portfolio of investments, programs, products and services.	

A. Component: Process		
Management Practice	**Example I&T Risk-specific Metrics**	
APO05.01 Determine the availability and sources of funds. Determine potential sources of funds, different funding options and the implications of the funding source on the investment return expectations.	a. Percent of funds dedicated to I&T risk management purposes b. Percent of sufficiently and timely funded I&T risk management initiatives	
I&T Risk-specific Activities (in addition to COBIT 2019 core-model activities)		**Capability Level**
1. Determine the availability of different funding options.		2
2. Build a solid business case emphasizing return on investment (ROI).		3
Related Guidance (Standards, Frameworks, Compliance Requirements)	**Detailed Reference**	
No related guidance for this management practice		
Management Practice	**Example I&T Risk-specific Metrics**	
APO05.02 Evaluate and select programs to fund. Based on requirements for the overall investment portfolio mix and the I&T strategic plan and road map, evaluate and prioritize program business cases and decide on investment proposals. Allocate funds and initiate programs.	a. Percent of I&T risk mitigation activities funded based on the formal cost/benefit analysis	
I&T Risk-specific Activities (in addition to COBIT 2019 core-model activities)		**Capability Level**
1. Confirm that the I&T risk universe covers key programs.		2
2. Determine the number of business objectives cascaded into actual projects.		
3. Prioritize I&T risk mitigation initiatives.		3
4. Use the I&T risk management procedures/methodology in the evaluation of programs.		
5. Ensure that selected programs comply with I&T risk requirements such as enterprise risk appetite/tolerance.		
6. Define methodologies to calculate the cost and benefit of risk mitigation activities.		4
Related Guidance (Standards, Frameworks, Compliance Requirements)	**Detailed Reference**	
PMBOK Guide Sixth Edition, 2017	Part 1: 1.2.3 Relationship of project, program, portfolio and operations management	

A. Component: Process *(cont.)*		
Management Practice	**Example I&T Risk-specific Metrics**	
APO05.03 Monitor, optimize and report on investment portfolio performance. On a regular basis, monitor and optimize the performance of the investment portfolio and individual programs throughout the entire investment life cycle. Ensure continuous follow-up on the alignment of the portfolio with I&T strategy.	a. ROI of selected I&T risk management programs b. Percent improvement in the overall level of I&T risk after investments	
I&T Risk-specific Activities (in addition to COBIT 2019 core-model activities)		**Capability Level**
1. Measure and report the achieved risk levels against planned levels.		3
2. Measure and report the ROI of the I&T risk management programs.		
3. Ensure that quality assurance activities are performed to monitor ROI achievement and ensure alignment with business objectives.		
Related Guidance (Standards, Frameworks, Compliance Requirements)	**Detailed Reference**	
No related guidance for this management practice		
Management Practice	**Example I&T Risk-specific Metrics**	
APO05.04 Maintain portfolios. Maintain portfolios of investment programs and projects, I&T products and services, and I&T assets.	a. Number of I&T products and services linked to business products and services	
I&T Risk-specific Activities (in addition to COBIT 2019 core-model activities)		**Capability Level**
1. Maintain an up-to-date information assets inventory.		2
2. Maintain an up-to-date service catalog.		
Related Guidance (Standards, Frameworks, Compliance Requirements)	**Detailed Reference**	
ITIL V3, 2011	Service Strategy, 4.2 Service portfolio management	
Management Practice	**Example I&T Risk-specific Metrics**	
APO05.05 Manage benefits achievement. Monitor the benefits of providing and maintaining appropriate I&T products, services and capabilities, based on the agreed and current business case.	a. Frequency of benefits reporting b. Number of KRIs related to the investment	
I&T Risk-specific Activities (in addition to COBIT 2019 core-model activities)		**Capability Level**
1. Assess and communicate improvements in the risk level.		2
Related Guidance (Standards, Frameworks, Compliance Requirements)	**Detailed Reference**	
No related guidance for this management practice		

C. Component: Information Flows and Items				
Management Practice	**I&T Risk-specific Inputs**		**I&T Risk-specific Outputs**	
	From	**Description**	**Description**	**To**
APO05.01 Determine the availability and sources of funds.	There are no I&T risk-specific inputs for this practice.		Business case for I&T risk initiatives	IRFA APO02.04
APO05.02 Evaluate and select programs to fund.	There are no I&T risk-specific inputs for this practice.		Risk action program business case	APO06.02
APO05.03 Monitor, optimize and report on investment portfolio performance.	Outside COBIT	Actual IT losses incurred as a consequence of materialized risk	Risk action program business case	Internal
APO05.04 Maintain portfolios.	There are no I&T risk-specific inputs or outputs for this practice.			
APO05.05 Manage benefits achievement.	There are no I&T risk-specific inputs for this practice.		Annual loss trends	Internal
Related Guidance (Standards, Frameworks, Compliance Requirements)	**Detailed Reference**			
No related guidance for this component				

Page intentionally left blank

Domain: Align, Plan and Organize Management Objective: APO06 — Managed Budget and Costs	Focus Area: I&T Risk
Description	
Manage the I&T-related financial activities in both the business and IT functions, covering budget, cost and benefit management and prioritization of spending through the use of formal budgeting practices and a fair and equitable system of allocating costs to the enterprise. Consult stakeholders to identify and control the total costs and benefits within the context of the I&T strategic and tactical plans. Initiate corrective action where needed.	
Purpose	
Foster a partnership between IT and enterprise stakeholders to enable the effective and efficient use of I&T-related resources and provide transparency and accountability of the cost and business value of solutions and services. Enable the enterprise to make informed decisions regarding the use of I&T solutions and services.	
I&T Risk Focus Area Relevance	
I&T risk management budgets are prepared, prioritized and maintained, and costs are managed.	

A. Component: Process		
Management Practice	**Example I&T Risk-specific Metrics**	
APO06.01 Manage finance and accounting. Establish and maintain a method to manage and account for all I&T-related costs, investments and depreciation as an integral part of enterprise financial systems and accounts. Report using the enterprise's financial measurement systems.	a. Percent of I&T risk estimations supported with sufficient financial information b. Number of business cases using I&T risk management inputs	
I&T Risk-specific Activities (in addition to COBIT 2019 core-model activities)		**Capability Level**
1. Ensure availability of sufficient financial data to enable estimation of I&T risk.		2
2. Promote/adopt the use of a quantitative risk management approach.		
3. Use the outputs of the quantitative risk assessment (monetary risk exposure, insurance premium, etc.) as inputs in the construction of business cases for investments/mitigation plans.		
4. Define the methodology to calculate the financial impact of identified I&T risk.		3
Related Guidance (Standards, Frameworks, Compliance Requirements)	**Detailed Reference**	
ITIL V3, 2011	Service Strategy, 4.3 Financial management for IT services	
Management Practice	**Example I&T Risk-specific Metrics**	
APO06.02 Prioritize resource allocation. Implement a decision-making process to prioritize the allocation of resources and establish rules for discretionary investments by individual business units. Include the potential use of external service providers and consider the buy, develop and rent options.	a. Number of risk maps b. Frequency of updates to risk maps c. Percent of alignment between I&T risk resources and high-priority I&T risk and control activities	
I&T Risk-specific Activities (in addition to COBIT 2019 core-model activities)		**Capability Level**
1. Establish a decision-making body for prioritizing business and I&T resources (including risk maps and external service providers) within the high-level budget allocations for I&T-enabled programs, I&T services and I&T assets.		2
Related Guidance (Standards, Frameworks, Compliance Requirements)	**Detailed Reference**	
No related guidance for this management practice		

A. Component: Process *(cont.)*		
Management Practice	**Example I&T Risk-specific Metrics**	
APO06.03 Create and maintain budgets. Prepare a budget reflecting investment priorities based on the portfolio of I&T-enabled programs and I&T services.	a. Number of I&T risk budget changes due to omissions or errors b. Number of budget overruns	
I&T Risk-specific Activities (in addition to COBIT 2019 core-model activities)		**Capability Level**
1. Account for the I&T risk function in budgets.		2
2. Monitor the I&T risk function budget and adjust it based on changing enterprise needs and financial considerations.		3
Related Guidance (Standards, Frameworks, Compliance Requirements)	**Detailed Reference**	
ISO/IEC 20000-1:2011(E)	6.4 Budgeting and accounting for services	
PMBOK Guide Sixth Edition, 2017	Part 1: 7. Project cost management	
Management Practice	**Example I&T Risk-specific Metrics**	
APO06.04 Model and allocate costs. Establish and use an I&T costing model based, for example, on the service definition. This approach ensures that allocation of costs for services is identifiable, measurable and predictable, and encourages the responsible use of resources, including those provided by service providers. Regularly review and benchmark the cost/chargeback model to maintain its relevance and appropriateness for evolving business and IT activities.	Additional I&T risk-specific metrics have not been identified for this practice. The COBIT 2019 core metrics are applicable.	
I&T Risk-specific Activities (in addition to COBIT 2019 core-model activities)		**Capability Level**
Additional I&T risk-specific activities have not been identified for this practice. The COBIT 2019 core activities are applicable.		N/A
Related Guidance (Standards, Frameworks, Compliance Requirements)	**Detailed Reference**	
No related guidance for this management practice		
Management Practice	**Example I&T Risk-specific Metrics**	
APO06.05 Manage costs. Implement a cost management process that compares actual costs against budget. Costs should be monitored and reported. Deviations from budget should be identified in a timely manner and their impact on enterprise processes and services assessed.	Additional I&T risk-specific metrics have not been identified for this practice. The COBIT 2019 core metrics are applicable.	
I&T Risk-specific Activities (in addition to COBIT 2019 core-model activities)		**Capability Level**
Additional I&T risk-specific activities have not been identified for this practice. The COBIT 2019 core activities are applicable.		N/A
Related Guidance (Standards, Frameworks, Compliance Requirements)	**Detailed Reference**	
No related guidance for this management practice		

C. Component: Information Flows and Items				
Management Practice	**I&T Risk-specific Inputs**		**I&T Risk-specific Outputs**	
AP006.01 Manage finance and accounting.	**From**	**Description**	**Description**	**To**
	IRFA AP001.01	Risk quantification approach (e.g., quantify risk impact)	Risk management plan	IRFA AP014.04
			Risk reserves	IRFA AP006.03
AP006.02 Prioritize resource allocation.	IRFA AP005.02	Risk action program business case	There are no I&T risk-specific outputs for this practice.	
AP006.03 Create and maintain budgets.	IRFA AP006.01	Risk reserves	There are no I&T risk-specific outputs for this practice.	
AP006.04 Model and allocate costs.	There are no I&T risk-specific inputs or outputs for this practice.			
AP006.05 Manage costs.	There are no I&T risk-specific inputs or outputs for this practice.			

Related Guidance (Standards, Frameworks, Compliance Requirements)	Detailed Reference
PMBOK Guide Sixth Edition, 2017	Part 1: 7. Project cost management: Inputs and Outputs

Page intentionally left blank

Domain: Align, Plan and Organize Management Objective: APO07 — Managed Human Resources	Focus Area: I&T Risk
Description	
Provide a structured approach to ensure optimal recruitment/acquisition, planning, evaluation and development of human resources (both internal and external).	
Purpose	
Optimize human resources capabilities to meet enterprise objectives.	
I&T Risk Focus Area Relevance	
Qualified I&T risk human resources are competent, skilled, managed and optimized, and available as required.	

A. Component: Process		
Management Practice	**Example I&T Risk-specific Metrics**	
APO07.01 Acquire and maintain adequate and appropriate staffing. Evaluate staffing requirements on a regular basis, or upon major changes to the enterprise, operational or IT environments, to ensure that the enterprise has sufficient human resources to support enterprise goals and objectives. Staffing includes both internal and external resources.	a. Average duration of I&T risk management position vacancies b. Percent of I&T risk posts vacant c. Rate of turnover in I&T risk management function	
I&T Risk-specific Activities (in addition to COBIT 2019 core-model activities)		**Capability Level**
1. Identify resource requirements for I&T risk management at both the business and I&T levels and in the context of competing business risk issues, resource limitations and objectives.		2
2. Allocate appropriate funds to fill risk staffing gaps and position the enterprise to take advantage of opportunities.		
3. Make risk/reward trade-offs in relation to organizational objectives.		3
4. Consider required skills (specifying how the risk management skills of managers and staff will be developed and maintained), documented processes and procedures for I&T risk management.		
5. Consider information systems and databases for managing I&T risk issues, budget and other resources for specific risk response activities, expectations from regulators, external auditors, etc.		
Related Guidance (Standards, Frameworks, Compliance Requirements)	**Detailed Reference**	
COSO Enterprise Risk Management, June 2017	6. Governance and Culture—Principle 5	
Skills Framework for the Information Age V6, 2015	SFIA and skills management—Acquire	
Management Practice	**Example I&T Risk-specific Metrics**	
APO07.02 Identify key IT personnel. Identify key IT personnel. Use knowledge capture (documentation), knowledge sharing, succession planning and staff backup to minimize reliance on a single individual performing a critical job function.	Additional I&T risk-specific metrics have not been identified for this practice. The COBIT 2019 core metrics are applicable.	
I&T Risk-specific Activities (in addition to COBIT 2019 core-model activities)		**Capability Level**
1. Identify and train a backup resource for key personnel.		2
2. Build a solid knowledge transfer mechanism.		3
3. Develop and maintain a succession plan for key risk personnel.		
Related Guidance (Standards, Frameworks, Compliance Requirements)	**Detailed Reference**	
CMMI Cybermaturity Platform, 2018	RI.RR Identification of Roles and Responsibilities	
Skills Framework for the Information Age V6, 2015	SFIA and skills management—Acquire	

A. Component: Process *(cont.)*		
Management Practice	**Example I&T Risk-specific Metrics**	
APO07.03 Maintain the skills and competencies of personnel. Define and manage the skills and competencies required of personnel. Regularly verify that personnel have the competencies to fulfill their roles on the basis of their education, training and/or experience. Verify that these competencies are being maintained, using qualification and certification programs where appropriate. Provide employees with ongoing learning and opportunities to maintain their knowledge, skills and competencies at a level required to achieve enterprise goals.	a. Qualification of staff in terms of certifications, education and years of experience b. Number of staff attending I&T risk awareness training c. Percent of scheduled I&T risk training programs provided	
I&T Risk-specific Activities (in addition to COBIT 2019 core-model activities)		**Capability Level**
1. Provide professional development training and programs on risk management.		2
2. Use certification to ensure a quality risk management professional skill set.		3
3. Establish appropriate enterprisewide education, training and awareness programs for risk.		
Related Guidance (Standards, Frameworks, Compliance Requirements)	**Detailed Reference**	
ISF, The Standard of Good Practice for Information Security 2016	PM2.3 Security Education/Training	
ISO/IEC 27001:2013/Cor.2:2015(E)	7.2 Competence	
National Institute of Standards and Technology Framework for Improving Critical Infrastructure Cybersecurity V1.1, April 2018	PR.AT Awareness and Training	
National Institute of Standards and Technology Special Publication 800-53, Revision 5 (Draft), August 2017	3.2 Awareness and training (AT-3, AT-4)	
Skills Framework for the Information Age V6, 2015	SFIA and skills management—Deploy	
The CIS Critical Security Controls for Effective Cyber Defense Version 6.1, August 2016	CSC 17: Security Skills Assessment and Appropriate Training to Fill Gaps	
Management Practice	**Example I&T Risk-specific Metrics**	
APO07.04 Assess and recognize/reward employee job performance. Conduct timely, regular performance evaluations against individual objectives derived from enterprise goals, established standards, specific job responsibilities, and the skills and competency framework. Implement a remuneration/recognition process that rewards successful attainment of performance goals.	Additional I&T risk-specific metrics have not been identified for this practice. The COBIT 2019 core metrics are applicable.	
I&T Risk-specific Activities (in addition to COBIT 2019 core-model activities)		**Capability Level**
Additional I&T risk-specific activities have not been identified for this practice. The COBIT 2019 core activities are applicable.		N/A
Related Guidance (Standards, Frameworks, Compliance Requirements)	**Detailed Reference**	
Skills Framework for the Information Age V6, 2015	SFIA and skills management—Develop	
Management Practice	**Example I&T Risk-specific Metrics**	
APO07.05 Plan and track the usage of IT and business human resources. Understand and track the current and future demand for business and IT human resources with responsibilities for enterprise I&T. Identify shortfalls and provide input into sourcing plans, enterprise and IT recruitment processes, and business and IT recruitment processes.	a. Number of I&T risk management staff allocated to each area relative to the I&T risk management plan	
I&T Risk-specific Activities (in addition to COBIT 2019 core-model activities)		**Capability Level**
1. Manage allocation of risk management staff according to the needs of the enterprise.		2
Related Guidance (Standards, Frameworks, Compliance Requirements)	**Detailed Reference**	
Skills Framework for the Information Age V6, 2015	SFIA and skills management—Assess; Reward	

A. Component: Process *(cont.)*		
Management Practice	**Example I&T Risk-specific Metrics**	
APO07.06 Manage contract staff. Ensure that consultants and contract personnel who support the enterprise with I&T skills know and comply with the organization's policies and meet agreed contractual requirements.	Additional I&T risk-specific metrics have not been identified for this practice. The COBIT 2019 core metrics are applicable.	
I&T Risk-specific Activities (in addition to COBIT 2019 core-model activities)		**Capability Level**
Additional I&T risk-specific activities have not been identified for this practice. The COBIT 2019 core activities are applicable.		N/A
Related Guidance (Standards, Frameworks, Compliance Requirements)	**Detailed Reference**	
Skills Framework for the Information Age V6, 2015	SFIA and skills management—Deploy	

C. Component: Information Flows and Items				
Management Practice	**I&T Risk-specific Inputs**		**I&T Risk-specific Outputs**	
	From	**Description**	**Description**	**To**
APO07.01 Acquire and maintain adequate and appropriate staffing.	IRFA APO2.06	Risk management plan	Risk function requirements for the staffing process	IRFA APO01.01
APO07.02 Identify key IT personnel.	There are no I&T risk-specific inputs or outputs for this practice.			
APO07.03 Maintain the skills and competencies of personnel.	There are no I&T risk-specific inputs for this practice.		Risk function training plan	IRFA APO07.04
			Risk awareness training curriculum	IRFA EDM03.03 IRFA APO14.01 IRFA BAI05.01 IRFA BAI05.02
APO07.04 Assess and recognize/reward employee job performance.	IRFA APO07.03	Risk function training plan	Risk function personnel evaluations	Internal
	Outside COBIT	HR policy		
APO07.05 Plan and track the usage of IT and business human resources.	Outside COBIT	Process resource requirements, budget allocations, personnel lists, skills	Resource performance tracking plan and indicators	Internal
APO07.06 Manage contract staff.	There are no I&T risk-specific inputs or outputs for this practice.			
Related Guidance (Standards, Frameworks, Compliance Requirements)	**Detailed Reference**			
PMBOK Guide Sixth Edition, 2017	Part 1: 9. Project resource management: Inputs and Outputs			

Page intentionally left blank

Domain: Align, Plan and Organize Management Objective: APO08 — Managed Relationships	Focus Area: I&T Risk
Description	
Manage relationships with business stakeholders in a formalized and transparent way that ensures mutual trust and a combined focus on achieving the strategic goals within the constraints of budgets and risk tolerance. Base relationships on open and transparent communication, a common language, and the willingness to take ownership and accountability for key decisions on both sides. Business and IT must work together to create successful enterprise outcomes in support of the enterprise objectives.	
Purpose	
Enable the right knowledge, skills and behaviors to create improved outcomes, increased confidence, mutual trust and effective use of resources that stimulate a productive relationship with business stakeholders.	
I&T Risk Focus Area Relevance	
I&T risk must maintain its integral and productive relationships with I&T and business stakeholders.	

A. Component: Process		
Management Practice	**Example I&T Risk-specific Metrics**	
APO08.01 Understand business expectations. Understand current business issues, objectives and expectations for I&T. Ensure that requirements are understood, managed and communicated, and their status agreed and approved.	a. Percent of business plan items linked to I&T risk	
I&T Risk-specific Activities (in addition to COBIT 2019 core-model activities)		**Capability Level**
1. Understand the business and how I&T risk enables/affects it.		2
Related Guidance (Standards, Frameworks, Compliance Requirements)	**Detailed Reference**	
No related guidance for this management practice		
Management Practice	**Example I&T Risk-specific Metrics**	
APO08.02 Align I&T strategy with business expectations and identify opportunities for IT to enhance the business. Align I&T strategies with current business objectives and expectations to enable IT to be a value-add partner for the business and a governance component for enhanced enterprise performance.	a. Percent of I&T risk management initiatives included in investment proposals b. Percent of I&T initiatives linked to business I&T risk management objectives	
I&T Risk-specific Activities (in addition to COBIT 2019 core-model activities)		**Capability Level**
1. Understand new technologies and how these can be applied innovatively and aligned to the enterprise's risk appetite.		3
2. Understand business risk and how I&T can help better manage it.		
Related Guidance (Standards, Frameworks, Compliance Requirements)	**Detailed Reference**	
ITIL V3, 2011	Service Strategy, 4.4 Demand management	

A. Component: Process *(cont.)*		
Management Practice	**Example I&T Risk-specific Metrics**	
APO08.03 Manage the business relationship. Manage the relationship between the IT service organization and its business partners. Ensure that relationship roles and responsibilities are defined and assigned, and communication is facilitated.	a. Number of I&T risk champions/correspondents per business unit	
I&T Risk-specific Activities (in addition to COBIT 2019 core-model activities)		**Capability Level**
1. Define the responsibilities for communication and escalation of I&T risk.		2
2. Identify partners (risk champions) who are both keen to adopt risk practices and sensitive to risk issues within business units.		
3. Develop a network of risk champions within the business units.		
4. Maintain a sustainable relationship with the risk champions.		
5. Establish clear roles and responsibilities for I&T risk.		
Related Guidance (Standards, Frameworks, Compliance Requirements)		
ISO/IEC 20000-1:2011(E)	7.1 Business relationship management	
ITIL V3, 2011	Service Strategy, 4.5 Business relationship management	
Management Practice	**Example I&T Risk-specific Metrics**	
APO08.04 Coordinate and communicate. Work with all relevant stakeholders and coordinate the end-to-end delivery of I&T services and solutions provided to the business.	a. Number of direct communications from the I&T risk function to various stakeholders b. Stakeholder satisfaction with I&T risk management activities and results, measured through stakeholder surveys	
I&T Risk-specific Activities (in addition to COBIT 2019 core-model activities)		**Capability Level**
1. Establish appropriate communication channels between the risk function and the business.		3
2. Establish the appropriate reporting and metrics regarding risk management.		4
Related Guidance (Standards, Frameworks, Compliance Requirements)	**Detailed Reference**	
No related guidance for this management practice		
Management Practice	**Example I&T Risk-specific Metrics**	
APO08.05 Provide input to the continual improvement of services. Continually improve and evolve I&T-enabled services and service delivery to the enterprise to align with changing enterprise objectives and technology requirements.	a. Number of I&T risk items used as inputs to action plans	
I&T Risk-specific Activities (in addition to COBIT 2019 core-model activities)		**Capability Level**
1. Monitor action plans to ensure that they are completed in accordance with risk analysis results.		5
2. Use risk analysis results as input for defining action plans and continuously improving the business.		
Related Guidance (Standards, Frameworks, Compliance Requirements)	**Detailed Reference**	
No related guidance for this management practice		

C. Component: Information Flows and Items				
Management Practice	**I&T Risk-specific Inputs**		**I&T Risk-specific Outputs**	
	From	**Description**	**Description**	**To**
APO08.01 Understand business expectations.	Outside COBIT	Business goals and objectives	Business strategy	IRFA APO08.02 IRFA APO08.03
APO08.02 Align I&T strategy with business expectations and identify opportunities for IT to enhance the business.	IRFA APO08.01	Business strategy	There are no I&T risk-specific outputs for this practice.	
APO08.03 Manage the business relationship.	IRFA APO08.01	Business strategy	Strategy to obtain stakeholder commitment	Internal
APO08.04 Coordinate and communicate.	Outside COBIT	Enterprise communication plan	Risk management communication strategy	Internal
APO08.05 Provide input to the continual improvement of services.	There are no I&T risk-specific inputs for this practice.		Integration of risk management in continual improvement processes	Internal
Related Guidance (Standards, Frameworks, Compliance Requirements)	**Detailed Reference**			
No related guidance for this component				

Page intentionally left blank

Domain: Align, Plan and Organize Management Objective: APO10 — Managed Vendors	Focus Area: I&T Risk
Description	
Manage I&T-related products and services provided by all types of vendors to meet enterprise requirements. This includes the search for and selection of vendors, management of relationships, management of contracts, and reviewing and monitoring of vendor performance and vendor ecosystem (including upstream supply chain) for effectiveness and compliance.	
Purpose	
Optimize available I&T capabilities to support the I&T strategy and road map, minimize the risk associated with nonperforming or noncompliant vendors, and ensure competitive pricing.	
I&T Risk Focus Area Relevance	
I&T risk relating to the enterprise's vendors is identified and appropriately managed.	

A. Component: Process		
Management Practice	**Example I&T Risk-specific Metrics**	
APO10.01 Identify and evaluate vendor relationships and contracts. Continuously search for and identify vendors and categorize them into type, significance and criticality. Establish criteria to evaluate vendors and contracts. Review the overall portfolio of existing and alternative vendors and contracts.	a. Percent of vendors that have undergone a formal I&T risk assessment b. Percent of contracted vendors without an I&T risk assessment	
I&T Risk-specific Activities (in addition to COBIT 2019 core-model activities)		**Capability Level**
1. Conduct information risk assessments and define the information risk profile.		3
2. Define the supplier relationship and requirements based on the information risk profile.		
Related Guidance (Standards, Frameworks, Compliance Requirements)	**Detailed Reference**	
No related guidance for this management practice		
Management Practice	**Example I&T Risk-specific Metrics**	
APO10.02 Select vendors. Select suppliers according to a fair and formal practice to ensure a viable best fit based on specified requirements. Requirements should be optimized with input from potential suppliers.	a. Number of vendors under contract with delivery delays b. Number of vendors under contract with poor quality delivery	
I&T Risk-specific Activities (in addition to COBIT 2019 core-model activities)		**Capability Level**
Additional I&T risk-specific activities have not been identified for this practice. The COBIT 2019 core activities are applicable.		N/A
Related Guidance (Standards, Frameworks, Compliance Requirements)	**Detailed Reference**	
No related guidance for this management practice		

A. Component: Process *(cont.)*		
Management Practice	**Example I&T Risk-specific Metrics**	
APO10.03 Manage vendor relationships and contracts. Formalize and manage the supplier relationship for each supplier. Manage, maintain and monitor contracts and service delivery. Ensure that new or changed contracts conform to enterprise standards and legal and regulatory requirements. Deal with contractual disputes.	a. Number of service level agreement (SLA) failures	
I&T Risk-specific Activities (in addition to COBIT 2019 core-model activities)		**Capability Level**
Additional I&T risk-specific activities have not been identified for this practice. The COBIT 2019 core activities are applicable.		N/A
Related Guidance (Standards, Frameworks, Compliance Requirements)	**Detailed Reference**	
ISO/IEC 20000-1:2011(E)	7.2 Supplier management	
ITIL V3, 2011	Service Design, 4.8 Supplier Management	
Management Practice	**Example I&T Risk-specific Metrics**	
APO10.04 Manage vendor risk. Identify and manage risk relating to vendors' ability to continually provide secure, efficient and effective service delivery. This also includes the subcontractors or upstream vendors that are relevant in the service delivery of the direct vendor.	a. Number of vendors not meeting service criteria	
I&T Risk-specific Activities (in addition to COBIT 2019 core-model activities)		**Capability Level**
Additional I&T risk-specific activities have not been identified for this practice The COBIT 2019 core activities are applicable.		N/A
Related Guidance (Standards, Frameworks, Compliance Requirements)	**Detailed Reference**	
CMMI Cybermaturity Platform, 2018	RM.MP Manage External Participation	
ISF, The Standard of Good Practice for Information Security 2016	SC1.1 External Supplier Management Process	
ISO/IEC 27002:2013/Cor.2:2015(E)	15. Supplier relationships	
National Institute of Standards and Technology Framework for Improving Critical Infrastructure Cybersecurity v1.1, April 2018	ID.SC Supply Chain Risk Management	
Management Practice	**Example I&T Risk-specific Metrics**	
APO10.05 Monitor vendor performance and compliance. Periodically review overall vendor performance, compliance to contract requirements and value for money. Address identified issues.	a. Percent of noncompliant vendors b. Percent of contract breaches c. Number of outstanding issues by vendor, risk rank or risk category	
I&T Risk-specific Activities (in addition to COBIT 2019 core-model activities)		**Capability Level**
1. Ensure that risk issues are aged to determine which have not yet been remediated or those whose planned remediations are delayed.		3
Related Guidance (Standards, Frameworks, Compliance Requirements)	**Detailed Reference**	
No related guidance for this management practice		

C. Component: Information Flows and Items				
Management Practice	**I&T Risk-specific Inputs**		**I&T Risk-specific Outputs**	
	From	**Description**	**Description**	**To**
APO10.01 Identify and evaluate vendor relationships and contracts.	There are no I&T risk-specific inputs for this practice.		Vendor risk management plan	IRFA APO10.04 IRFA BAI03.04 IRFA DSS01.02
APO10.02 Select vendors.	There are no I&T risk-specific inputs or outputs for this practice.			
APO10.03 Manage vendor relationships and contracts.	There are no I&T risk-specific inputs or outputs for this practice.			
APO10.04 Manage vendor risk.	There are no I&T risk-specific inputs or outputs for this practice.			
APO10.05 Monitor vendor performance and compliance.	IRFA MEA03.01	Report of vendor compliance breaches	Vendor remediation plan	Internal
Related Guidance (Standards, Frameworks, Compliance Requirements)	**Detailed Reference**			
No related guidance for this component				

Page intentionally left blank

Domain: Align, Plan and Organize Management Objective: APO11 — Managed Quality	Focus Area: I&T Risk
Description	
Define and communicate quality requirements in all processes, procedures and related enterprise outcomes. Enable controls, ongoing monitoring, and the use of proven practices and standards in continuous improvement and efficiency efforts.	
Purpose	
Ensure consistent delivery of technology solutions and services to meet the quality requirements of the enterprise and satisfy stakeholder needs.	
I&T Risk Focus Area Relevance	
The quality of I&T risk management services needs to be established and maintained to meet the quality requirements of the enterprise and its stakeholders.	

A. Component: Process		
Management Practice	**Example I&T Risk-specific Metrics**	
APO11.01 Establish a quality management system (QMS). Establish and maintain a quality management system (QMS) that provides a standard, formal and continuous approach to quality management of information. The QMS should enable technology and business processes to align with business requirements and enterprise quality management.	a. Stakeholder satisfaction with I&T risk management activities and results, measured through stakeholder surveys	
I&T Risk-specific Activities (in addition to COBIT 2019 core-model activities)		**Capability Level**
1. Manage quality to mitigate risk with associated I&T processes and services.		3
Related Guidance (Standards, Frameworks, Compliance Requirements)	**Detailed Reference**	
PMBOK Guide Sixth Edition, 2017	Part 1: 8.1 Plan quality management	
Management Practice	**Example I&T Risk-specific Metrics**	
APO11.02 Focus quality management on customers. Focus quality management on customers by determining their requirements and ensuring integration in quality management practices.	Additional I&T risk-specific metrics have not been identified for this practice. The COBIT 2019 core metrics are applicable.	
I&T Risk-specific Activities (in addition to COBIT 2019 core-model activities)		**Capability Level**
Additional I&T risk-specific activities have not been identified for this practice. The COBIT 2019 core activities are applicable.		N/A
Related Guidance (Standards, Frameworks, Compliance Requirements)	**Detailed Reference**	
No related guidance for this management practice		
Management Practice	**Example I&T Risk-specific Metrics**	
APO11.03 Manage quality standards, practices and procedures and integrate quality management into key processes and solutions. Identify and maintain standards, procedures and practices for key processes to guide the enterprise in meeting the intent of the agreed quality management standards (QMS). This activity should align with I&T control framework requirements. Consider certification for key processes, organizational units, products or services.	a. Number of I&T risk management practices aligned with the QMS	
I&T Risk-specific Activities (in addition to COBIT 2019 core-model activities)		**Capability Level**
1. Align the risk management practices with the quality management system (QMS).		3
2. Consider the benefits and cost of external quality reviews.		
Related Guidance (Standards, Frameworks, Compliance Requirements)	**Detailed Reference**	
PMBOK Guide Sixth Edition, 2017	Part 1: 8.2 Manage quality	

A. Component: Process *(cont.)*		
Management Practice	**Example I&T Risk-specific Metrics**	
APO11.04 Perform quality monitoring, control and reviews. Monitor the quality of processes and services on an ongoing basis, in line with quality management standards. Define, plan and implement measurements to monitor customer satisfaction with quality as well as the value provided by the quality management system (QMS). The information gathered should be used by the process owner to improve quality.	a. Number of I&T risk management quality metrics that are below quality baselines b. Frequency of I&T risk reporting c. Number of corrective actions needed in I&T risk function d. Number of I&T risk management quality metrics in place	
I&T Risk-specific Activities (in addition to COBIT 2019 core-model activities)		**Capability Level**
1. Define risk management quality metrics to measure the achievement of risk management requirements and the efficient functioning of risk controls.		3
2. Monitor the risk management quality metrics.		4
3. Validate I&T risk management information, calculations and metrics (KRIs) for accuracy and completeness.		
4. Take corrective action to address quality issues in the risk function.		5
Related Guidance (Standards, Frameworks, Compliance Requirements)	**Detailed Reference**	
PMBOK Guide Sixth Edition, 2017	Part 1: 8.3 Control quality	
Management Practice	**Example I&T Risk-specific Metrics**	
APO11.05 Maintain continuous improvement. Maintain and regularly communicate an overall quality plan that promotes continuous improvement. The plan should define the need for, and benefits of, continuous improvement. Collect and analyze data about the quality management system (QMS) and improve its effectiveness. Correct nonconformities to prevent recurrence.	a. Number of root cause analyses for I&T risk management issues performed b. Number of corrective practices applied with respect to I&T risk issues	
I&T Risk-specific Activities (in addition to COBIT 2019 core-model activities)		**Capability Level**
1. Identify, document and communicate root causes for risk management issues with quality metrics.		4
2. Apply corrective practices to remediate quality issues.		5
Related Guidance (Standards, Frameworks, Compliance Requirements)	**Detailed Reference**	
National Institute of Standards and Technology Framework for Improving Critical Infrastructure Cybersecurity v1.1, April 2018	DE.DP Detection Processes	

C. Component: Information Flows and Items				
Management Practice	**I&T Risk-specific Inputs**		**I&T Risk-specific Outputs**	
	From	**Description**	**Description**	**To**
APO11.01 Establish a quality management system (QMS).	IRFA APO02.06	Risk management plan	Relevant risk function best practices and standards	IRFA APO11.03 IRFA MEA01.01
APO11.02 Focus quality management on customers.	IRFA APO11.03	Risk function quality standards	There are no I&T risk-specific outputs for this practice.	
APO11.03 Manage quality standards, practices and procedures and integrate quality management into key processes and solutions.	IRFA APO11.01	Relevant risk function best practices and standards	Risk function quality standards	IRFA APO11.02 IRFA APO11.04
APO11.04 Perform quality monitoring, control and reviews.	IRFA APO11.03	Risk function quality standards	Risk function quality metrics implemented in line with best practices	IRFA APO11.05 IRFA MEA01.01
APO11.05 Maintain continuous improvement.	IRFA APO11.04	Risk function quality metrics implemented in line with best practices	Documented root causes for risk management issues with quality metrics	Internal

Related Guidance (Standards, Frameworks, Compliance Requirements)	Detailed Reference
PMBOK Guide Sixth Edition, 2017	Part 1: 8. Project quality management: Inputs and Outputs

Page intentionally left blank

Domain: Align, Plan and Organize Management Objective: APO14 – Managed Data	Focus Area: I&T Risk
Description	
Achieve and sustain effective management of the enterprise data assets across the data life cycle, from creation through delivery, maintenance and archiving.	
Purpose	
Ensure effective utilization of the critical data assets to achieve enterprise goals and objectives.	
I&T Risk Focus Area Relevance	
I&T risk management data needs to be secure and appropriately managed.	

A. Component: Process

Management Practice	Example I&T Risk-specific Metrics
APO14.01 Define and communicate the organization's data management strategy and roles and responsibilities. Define how to manage and improve the organization's data assets, in line with enterprise strategy and objectives. Communicate the data management strategy to all stakeholders. Assign roles and responsibilities to ensure that corporate data are managed as critical assets and the data management strategy is implemented and maintained in an effective and sustainable manner.	a. Frequency of review of the I&T risk data strategy b. Number of staff undergoing awareness training on the I&T risk of sharing information

I&T Risk-specific Activities (in addition to COBIT 2019 core-model activities)	Capability Level
1. Establish appropriate ownership and responsibility for managing information and data privacy to ensure compliance with legal and regulatory requirements and reduce the risk of misuse of critical information.	2
2. Define the risk data strategy and align it with business strategies and the enterprise's overall objectives and risk appetite.	3
3. Provide awareness training on the risk of sharing information.	

Related Guidance (Standards, Frameworks, Compliance Requirements)	Detailed Reference
CMMI Data Management Maturity Model, 2014	Data Management Strategy - Data Management Strategy; Data Governance—Governance Management
ITIL V3, 2011	Service Design, 5.2 Management of Data and Information
The CIS Critical Security Controls for Effective Cyber Defense Version 6.1, August 2016	CSC 13: Data Protection

Management Practice	Example I&T Risk-specific Metrics
APO14.02 Define and maintain a consistent business glossary. Create, approve, update and promote consistent business terms and definitions to foster shared data usage across the organization.	a. Frequency of review of the I&T risk glossary

I&T Risk-specific Activities (in addition to COBIT 2019 core-model activities)	Capability Level
1. Ensure that standard risk terms are readily available and communicated to relevant stakeholders.	2
2. Ensure that each risk term added to the risk glossary has a unique name and unique definition.	
3. Use standard industry risk terms and definitions, as appropriate, in the risk glossary.	
4. Periodically review risk terms in the business glossary to ensure accuracy; update as needed.	
5. Integrate the risk glossary into the organization's metadata repository, with appropriate access permissions.	3

Related Guidance (Standards, Frameworks, Compliance Requirements)	Detailed Reference
CMMI Data Management Maturity Model, 2014	Data Governance - Business Glossary
ISF, The Standard of Good Practice for Information Security 2016	IM1.1 Information Classification and Handling

A. Component: Process *(cont.)*		
Management Practice	**Example I&T Risk-specific Metrics**	
APO14.03 Establish the processes and infrastructure for metadata management. Establish the processes and infrastructure for specifying and extending metadata about the organization's data assets, fostering and supporting data sharing, ensuring compliant use of data, improving responsiveness to business changes and reducing data-related risk.	a. Frequency of I&T risk glossary updates	
I&T Risk-specific Activities (in addition to COBIT 2019 core-model activities)		**Capability Level**
Additional I&T risk-specific activities have not been identified for this practice. The COBIT 2019 core activities are applicable.		N/A
Related Guidance (Standards, Frameworks, Compliance Requirements)	**Detailed Reference**	
CMMI Data Management Maturity Model, 2014	Data Governance—Metadata Management	
ISO/IEC 27002:2013/Cor.2:2015(E)	8.2 Information classification	
Management Practice	**Example I&T Risk-specific Metrics**	
APO14.04 Define a data quality strategy. Define an integrated, organizationwide strategy to achieve and maintain the level of data quality (such as complexity, integrity, accuracy, completeness, validity, traceability and timeliness) required to support the business goals and objectives.	a. Frequency and accuracy of I&T risk data quality reports	
I&T Risk-specific Activities (in addition to COBIT 2019 core-model activities)		**Capability Level**
1. Define a strategy for risk data quality in collaboration with all relevant risk stakeholders and approved by executive management. The strategy should facilitate the response and appropriate mitigation actions when necessary.		3
2. Ensure that the risk data quality strategy is followed across the organization and is accompanied by corresponding policies, processes and guidelines		
3. Develop, monitor and maintain a sequence plan for risk data quality improvement efforts across the organization.		
4. Systematically collect stakeholders' reports of risk issues. Include their expectations for improving risk data quality in the risk data quality strategy. Measure and monitor the results of the reported expectations.		4
Related Guidance (Standards, Frameworks, Compliance Requirements)	**Detailed Reference**	
CMMI Cybermaturity Platform, 2018	DP.DR Safeguard Data at Rest; DP.DT Safeguard Data in Transit; DP.IP Integrity and Data Leak Prevention	
CMMI Data Management Maturity Model, 2014	Data Quality - Data Quality Strategy	
Management Practice	**Example I&T Risk-specific Metrics**	
APO14.05 Establish data profiling methodologies, processes and tools. Implement standardized data profiling methodologies, processes, practices, tools and templates that can be applied across multiple data repositories and data stores.	Additional I&T risk-specific metrics have not been identified for this practice. The COBIT 2019 core metrics are applicable.	
I&T Risk-specific Activities (in addition to COBIT 2019 core-model activities)		**Capability Level**
Additional I&T risk-specific activities have not been identified for this practice. The COBIT 2019 core activities are applicable.		N/A
Related Guidance (Standards, Frameworks, Compliance Requirements)	**Detailed Reference**	
CMMI Data Management Maturity Model, 2014	Data Quality—Data Profiling	
National Institute of Standards and Technology Special Publication 800-53, Revision 5, August 2017	3.20 System and information integrity (SI-1)	

A. Component: Process *(cont.)*		
Management Practice	**Example I&T Risk-specific Metrics**	
APO14.06 Ensure a data quality assessment approach. Provide a systematic approach to measure and evaluate data quality according to processes and techniques, and against data quality rules.	a. Number of I&T risk quality assessments b. Number of errors in I&T risk data not yet remediated	
I&T Risk-specific Activities (in addition to COBIT 2019 core-model activities)		**Capability Level**
1. Periodically conduct risk data quality assessments, according to an approved frequency per the risk data quality assessment policy.		4
2. Include recommendations for remediation, with supporting rationale, in the risk data quality assessment results.		
Related Guidance (Standards, Frameworks, Compliance Requirements)	**Detailed Reference**	
CMMI Data Management Maturity Model, 2014	Data Quality—Data Quality Assessment	
Management Practice	**Example I&T Risk-specific Metrics**	
APO14.07 Define the data cleansing approach. Define the mechanisms, rules, processes, and methods to validate and correct data according to predefined business rules.	Additional I&T risk-specific metrics have not been identified for this practice. The COBIT 2019 core metrics are applicable.	
I&T Risk-specific Activities (in addition to COBIT 2019 core-model activities)		**Capability Level**
1. Determine the number of issues identified through random checks.		3
Related Guidance (Standards, Frameworks, Compliance Requirements)	**Detailed Reference**	
CMMI Data Management Maturity Model, 2014	Data Quality—Data Cleansing	
Management Practice	**Example I&T Risk-specific Metrics**	
APO14.08 Manage the life cycle of data assets. Ensure that the organization understands, maps, inventories and controls its data flows through business processes over the data life cycle, from creation or acquisition to retirement.	a. Number of I&T risk considerations contained in the enterprise information life cycle b. Number of data sets or data practices that do not comply with applicable laws and regulations	
I&T Risk-specific Activities (in addition to COBIT 2019 core-model activities)		**Capability Level**
1. Incorporate information risk considerations into the enterprise information life cycle.		3
2. Ensure systematic and structured security of information and data during their life cycle, in both digital and physical form.		
3. Ensure compliance with legal and regulatory requirements to preserve the confidentiality, integrity and availability of information and data and reduce their (potential) risks.		
Related Guidance (Standards, Frameworks, Compliance Requirements)	**Detailed Reference**	
CMMI Data Management Maturity Model, 2014	Data Operations—Data Lifecycle Management	
Management Practice	**Example I&T Risk-specific Metrics**	
APO14.09 Support data archiving and retention. Ensure that data maintenance satisfies organizational and regulatory requirements for availability of historical data. Ensure that legal and regulatory requirements for data archiving and retention are met.	a. Percent of historical I&T risk data that cannot be accessed b. Percent of historical I&T risk data that cannot be retrieved	
I&T Risk-specific Activities (in addition to COBIT 2019 core-model activities)		**Capability Level**
1. Ensure that policies mandate management of I&T risk data history, including retention, destruction and audit trail requirements.		2
2. Ensure the existence of a defined method that guarantees accessibility to the historical I&T risk data necessary to support business needs.		
3. Use policies and processes to control access, transmittal and modifications to historical and archived I&T risk data.		
Related Guidance (Standards, Frameworks, Compliance Requirements)	**Detailed Reference**	
CMMI Data Management Maturity Model, 2014	Platform and Architecture—Historical Data, Retention and Archiving	

A. Component: Process *(cont.)*		
Management Practice	**Example I&T Risk-specific Metrics**	
APO14.10 Manage data backup and restore arrangements. Manage availability of critical data to ensure operational continuity.	a. Percent of data that does not meet requirements for on-site and offsite storage b. Frequency of testing of backups on I&T risk data	
I&T Risk-specific Activities (in addition to COBIT 2019 core-model activities)		**Capability Level**
1. Define a schedule to ensure correct backup of all critical I&T risk data, taking into account frequency, type of backup, security, privacy and other criteria.		2
2. Define requirements for on-site and off-site storage of backup I&T risk data, taking into account volume, capacity and retention period, in alignment with business requirements.		
3. Establish a testing schedule for backup I&T risk data. Ensure that the I&T risk data can be restored correctly without drastically impacting business processes.		
Related Guidance (Standards, Frameworks, Compliance Requirements)	**Detailed Reference**	
The CIS Critical Security Controls for Effective Cyber Defense Version 6.1, August 2016	CSC 10: Data Recovery Capability	

C. Component: Information Flows and Items				
Management Practice	**I&T Risk-specific Inputs**		**I&T Risk-specific Outputs**	
	From	**Description**	**Description**	**To**
APO14.01 Define and communicate the organization's data management strategy and roles and responsibilities.	IRFA APO01.01	Data, process and control owners	Risk policies	IRFA APO14.09 IRFA APO14.10
	IRFA APO07.03	Risk awareness training curriculum		
APO14.02 Define and maintain a consistent business glossary.	There are no I&T risk-specific inputs for this practice.		IT risk glossary	Internal
APO14.03 Establish the processes and infrastructure for metadata management.	There are no I&T risk-specific inputs or outputs for this practice.			
APO14.04 Define a data quality strategy.	There are no I&T risk-specific inputs for this practice.		Risk data quality management policy	IRFA DSS06.02
APO14.05 Establish data profiling methodologies, processes and tools.	There are no I&T risk-specific inputs or outputs for this practice.			
APO14.06 Ensure a data quality assessment approach.	There are no I&T risk-specific inputs or outputs for this practice.			
APO14.07 Define the data cleansing approach.	There are no I&T risk-specific inputs or outputs for this practice.			
APO14.08 Manage the life cycle of data assets.	IRFA MEA03.01	Legally required risk management plan	There are no I&T risk-specific outputs for this practice.	
APO14.09 Support data archiving and retention.	IRFA APO14.01	Risk data policies	There are no I&T risk-specific outputs for this practice.	
APO14.10 Manage data backup and restore arrangements.	IRFA APO14.01	Risk data policies	There are no I&T risk-specific outputs for this practice.	
Related Guidance (Standards, Frameworks, Compliance Requirements)	**Detailed Reference**			
No related guidance for this component				

4.3 Build, Acquire and Implement (BAI)

Page intentionally left blank

Domain: Build, Acquire and Implement Management Objective: BAI03 – Managed Solutions Identification and Build	Focus Area: I&T Risk
Description	
Establish and maintain identified products and services (technology, business processes and workflows) in line with enterprise requirements covering design, development, procurement/sourcing and partnering with vendors. Manage configuration, test preparation, testing, requirements management and maintenance of business processes, applications, information/data, infrastructure and services.	
Purpose	
Ensure agile and scalable delivery of digital products and services. Establish timely and cost-effective solutions (technology, business processes and workflows) capable of supporting enterprise strategic and operational objectives.	
I&T Risk Focus Area Relevance	
I&T risk products, services, processes and solutions need to be designed, developed and acquired according to business requirements.	

A. Component: Process		
Management Practice	**Example I&T Risk-specific Metrics**	
BAI03.01 Design high-level solutions. Develop and document high-level designs for the solution in terms of technology, business processes and workflows. Use agreed and appropriate phased or rapid Agile development techniques. Ensure alignment with the I&T strategy and enterprise architecture. Reassess and update the designs when significant issues occur during detailed design or building phases, or as the solution evolves. Apply a user-centric approach; ensure that stakeholders actively participate in the design and approve each version.	a. Percent of stakeholder participation in the I&T risk identification and classification processes b. Number of internal/external audit findings	
I&T Risk-specific Activities (in addition to COBIT 2019 core-model activities)		**Capability Level**
1. Ensure that design is in alignment with enterprise risk tolerance.		3
2. Ensure that I&T risk management requirements are taken into account when designing appropriate solutions.		
Related Guidance (Standards, Frameworks, Compliance Requirements)	**Detailed Reference**	
ISF, The Standard of Good Practice for Information Security 2016	SD2.2 System Design	
Management Practice	**Example I&T Risk-specific Metrics**	
BAI03.02 Design detailed solution components. Develop, document and elaborate detailed designs progressively. Use agreed and appropriate phased or rapid Agile development techniques, addressing all components (business processes and related automated and manual controls, supporting I&T applications, infrastructure services and technology products, and partners/suppliers). Ensure that the detailed design includes internal and external service level agreements (SLAs) and operational level agreements (OLAs).	a. Number of I&T risk response strategies needing to be redefined due to incompatibility with the components design b. Percent of designed solutions subject to I&T risk assessment	
I&T Risk-specific Activities (in addition to COBIT 2019 core-model activities)		**Capability Level**
1. Ensure that all high-level risk response strategies have been translated to the detailed component design.		3
2. Proactively identify, analyze and assess I&T risk issues the solution may encompass/bring.		
3. Address critical risk issues by adopting a mitigation/corrective actions plan, in accordance with the enterprise's risk appetite/tolerance.		
Related Guidance (Standards, Frameworks, Compliance Requirements)	**Detailed Reference**	
ISF, The Standard of Good Practice for Information Security 2016	SD2.2 System Design	

A. Component: Process *(cont.)*	
Management Practice	**Example I&T Risk-specific Metrics**
BAI03.03 Develop solution components. Develop solution components progressively in a separate environment, in accordance with detailed designs following standards and requirements for development and documentation, quality assurance (QA), and approval. Ensure that all control requirements in the business processes, supporting I&T applications and infrastructure services, services and technology products, and partner/vendor services are addressed.	a. Number of I&T risk response strategies needing to be redefined due to incompatibility with the developed components b. Percent of the mitigation actions implemented by the developed solution

I&T Risk-specific Activities (in addition to COBIT 2019 core-model activities)	Capability Level
1. Ensure that all detailed risk response strategies have been successfully implemented into the developed components.	3
2. Ensure that mitigations actions are implemented in accordance with the mitigation plan adopted during the detailed design phase.	
3. Ensure proper communication by establishing a communication unit within the implementation team.	

Related Guidance (Standards, Frameworks, Compliance Requirements)	Detailed Reference
ISF, The Standard of Good Practice for Information Security 2016	SD1.2 System Development Environments
ISO/IEC 27002:2013/Cor.2:2015(E)	14.2 Security in development and support processes
ITIL V3, 2011	Service Strategy, 5.5 IT service strategy and application development
National Institute of Standards and Technology Special Publication 800-53, Revision 5 (Draft), August 2017	3.18 System and services acquisition (SA-3)

Management Practice	Example I&T Risk-specific Metrics
BAI03.04 Procure solution components. Procure solution components, based on the acquisition plan, in accordance with requirements and detailed designs, architecture principles and standards, and the enterprise's overall procurement and contract procedures, QA requirements, and approval standards. Ensure that all legal and contractual requirements are identified and addressed by the vendor.	a. Percent of stakeholder participation in the I&T risk identification and classification processes b. Percent of the mitigation/corrective actions implemented by the acquired solution

I&T Risk-specific Activities (in addition to COBIT 2019 core-model activities)	Capability Level
1. Ensure that mitigations/corrective actions are in place, in accordance with the mitigation plan adopted during the detailed design phase.	3

Related Guidance (Standards, Frameworks, Compliance Requirements)	Detailed Reference
ISF, The Standard of Good Practice for Information Security 2016	SD2.3 Software Acquisition
National Institute of Standards and Technology Framework for Improving Critical Infrastructure Cybersecurity v1.1, April 2018	3.4 Buying Decisions
National Institute of Standards and Technology Special Publication 800-53, Revision 5 (Draft), August 2017	3.18 System and services acquisition (SA-4)

A. Component: Process *(cont.)*		
Management Practice	**Example I&T Risk-specific Metrics**	
BAI03.05 Build solutions. Install and configure solutions and integrate with business process activities. During configuration and integration of hardware and infrastructure software, implement control, security, privacy and auditability measures to protect resources and ensure availability and data integrity. Update the product or services catalogue to reflect the new solutions.	a. Number of new I&T risk issues introduced with the solution	
I&T Risk-specific Activities (in addition to COBIT 2019 core-model activities)		**Capability Level**
1. Ensure the participation of information security and risk management.		2
2. Ensure the management of new risk emerging from the new solution.		3
Related Guidance (Standards, Frameworks, Compliance Requirements)	**Detailed Reference**	
HITRUST CSF version 9, September 2017	10.05 Security in Development & Support Processes	
ISF, The Standard of Good Practice for Information Security 2016	SD2.4 System Build	
Management Practice	**Example I&T Risk-specific Metrics**	
BAI03.06 Perform quality assurance (QA). Develop, resource and execute a QA plan aligned with the QMS to obtain the quality specified in the requirements definition and in the enterprise's quality policies and procedures.	a. Number of solutions assessed to be in excess of the enterprise's risk tolerance	
I&T Risk-specific Activities (in addition to COBIT 2019 core-model activities)		**Capability Level**
1. Periodically monitor that solution controls are within risk tolerance.		4
Related Guidance (Standards, Frameworks, Compliance Requirements)	**Detailed Reference**	
ISF, The Standard of Good Practice for Information Security 2016	SD1.3 Quality Assurance	
Management Practice	**Example I&T Risk-specific Metrics**	
BAI03.07 Prepare for solution testing. Establish a test plan and required environments to test the individual and integrated solution components. Include the business processes and supporting services, applications and infrastructure.	a. Percent of test cases mapped to stakeholder requirements	
I&T Risk-specific Activities (in addition to COBIT 2019 core-model activities)		**Capability Level**
Additional I&T risk-specific activities have not been identified for this practice. The COBIT 2019 core activities are applicable.		N/A
Related Guidance (Standards, Frameworks, Compliance Requirements)	**Detailed Reference**	
CMMI Cybermaturity Platform, 2018	AD.DE Safeguard Development Environment	
National Institute of Standards and Technology Special Publication 800-53, Revision 5 (Draft), August 2017	3.10 Maintenance (MA-2, MA-3)	

A. Component: Process *(cont.)*		
Management Practice	**Example I&T Risk-specific Metrics**	
BAI03.08 Execute solution testing. During development, execute testing continually (including control testing), in accordance with the defined test plan and development practices in the appropriate environment. Engage business process owners and end users in the test team. Identify, log and prioritize errors and issues identified during testing.	a. Percent of the solutions tested with critical controls deficiencies	
I&T Risk-specific Activities (in addition to COBIT 2019 core-model activities)		**Capability Level**
Additional I&T risk-specific activities have not been identified for this practice. The COBIT 2019 core activities are applicable.		N/A
Related Guidance (Standards, Frameworks, Compliance Requirements)	**Detailed Reference**	
CMMI Cybermaturity Platform, 2018	AD.ST Secure Development Testing	
ISF, The Standard of Good Practice for Information Security 2016	SD2.5 System Testing; SD2.6 Security Testing	
National Institute of Standards and Technology Special Publication 800-53, Revision 5 (Draft), August 2017	3.18 System and services acquisition (SA-11)	
Management Practice	**Example I&T Risk-specific Metrics**	
BAI03.09 Manage changes to requirements. Track the status of individual requirements (including all rejected requirements) throughout the project life cycle. Manage the approval of changes to requirements.	a. Number of new I&T risk issues introduced with changes to requirements b. Number of rejected requirements	
I&T Risk-specific Activities (in addition to COBIT 2019 core-model activities)		**Capability Level**
1. Assess the impact of all solution changes on defined risk response strategies.		3
Related Guidance (Standards, Frameworks, Compliance Requirements)	**Detailed Reference**	
ISF, The Standard of Good Practice for Information Security 2016	SD2.9 Post-implementation Review	
Management Practice	**Example I&T Risk-specific Metrics**	
BAI03.10 Maintain solutions. Develop and execute a plan for the maintenance of solution and infrastructure components. Include periodic reviews against business needs and operational requirements.	No I&T risk-specific metrics have been identified for this practice. The COBIT 2019 core metrics are applicable.	
I&T Risk-specific Activities (in addition to COBIT 2019 core-model activities)		**Capability Level**
1. Ensure that key performance indicators (KPIs) are measured for the solution.		2
Related Guidance (Standards, Frameworks, Compliance Requirements)	**Detailed Reference**	
ISO/IEC 27002:2013/Cor.2:2015(E)	14.3 Test data	
Management Practice	**Example I&T Risk-specific Metrics**	
BAI03.11 Define IT products and services and maintain the service portfolio. Define and agree on new or changed IT products or services and service level options. Document new or changed product and service definitions and service level options to be updated in the products and services portfolio.	a. Number of changed products, services and SLAs	
I&T Risk-specific Activities (in addition to COBIT 2019 core-model activities)		**Capability Level**
Additional I&T risk-specific activities have not been identified for this practice. The COBIT 2019 core activities are applicable.		N/A
Related Guidance (Standards, Frameworks, Compliance Requirements)	**Detailed Reference**	
No related guidance for this management practice		

A. Component: Process *(cont.)*		
Management Practice	**Example I&T Risk-specific Metrics**	
BAI03.12 Design solutions based on the defined development methodology. Design, develop and implement solutions with the appropriate development methodology (i.e., waterfall, Agile or bimodal I&T), in accordance with the overall strategy and requirements.	No I&T risk-specific metrics have been identified for this practice. The COBIT 2019 core metrics are applicable.	
I&T Risk-specific Activities (in addition to COBIT 2019 core-model activities)		**Capability Level**
Additional I&T risk-specific activities have not been identified for this practice. The COBIT 2019 core activities are applicable.		N/A
Related Guidance (Standards, Frameworks, Compliance Requirements)	**Detailed Reference**	
ISF, The Standard of Good Practice for Information Security 2016	SD1.1 System Development Methodology	

C. Component: Information Flows and Items				
Management Practice	**I&T Risk-specific Inputs**		**I&T Risk-specific Outputs**	
BAI03.01 Design high-level solutions.	**From**	**Description**	**Description**	**To**
	There are no I&T risk-specific inputs for this practice.		Risk section in design documentation	IRFA BAI03.05
BAI03.02 Design detailed solution components.	There are no I&T risk-specific inputs or outputs for this practice.			
BAI03.03 Develop solution components.	There are no I&T risk-specific inputs or outputs for this practice.			
BAI03.04 Procure solution components.	IRFA APO10.01	Vendor risk management plan	There are no I&T risk-specific outputs for this practice.	
BAI03.05 Build solutions.	IRFA BAI03.01	Risk section in design documentation	There are no I&T risk-specific outputs for this practice.	
BAI03.06 Perform quality assurance (QA).	There are no I&T risk-specific inputs or outputs for this practice.			
BAI03.07 Prepare for solution testing.	There are no I&T risk-specific inputs or outputs for this practice.			
BAI03.08 Execute solution testing.	There are no I&T risk-specific inputs or outputs for this practice.			
BAI03.09 Manage changes to requirements.	There are no I&T risk-specific inputs for this practice.		Record of risk associated with approved changes	IRFA DSS06.04
BAI03.10 Maintain solutions.	IRFA EDM03.01	Key risk indicators	There are no I&T risk-specific outputs for this practice.	
BAI03.11 Define IT products and services and maintain the service portfolio.	There are no I&T risk-specific inputs or outputs for this practice.			
BAI03.12 Design solutions based on the defined development methodology.	There are no I&T risk-specific inputs or outputs for this practice.			
Related Guidance (Standards, Frameworks, Compliance Requirements)	**Detailed Reference**			
No related guidance for this component				

Page intentionally left blank

Domain: Build, Acquire and Implement Management Objective: BAI05 – Managed Organizational Change	Focus Area: I&T Risk
Description	
Maximize the likelihood of successfully implementing sustainable enterprisewide organizational change quickly and with reduced risk. Cover the complete life cycle of the change and all affected stakeholders in the business and IT.	
Purpose	
Prepare and commit stakeholders for business change and reduce the risk of failure.	
I&T Risk Focus Area Relevance	
An appropriate I&T risk culture, with attention for both negative consequences and opportunities and benefits is created, enabled, driven and maintained through the organizational change processes and programs.	

A. Component: Process

Management Practice	Example I&T Risk-specific Metrics
BAI05.01 Establish the desire to change. Understand the scope and impact of the desired change. Assess stakeholder readiness and willingness to change. Identify actions that will motivate stakeholder acceptance and participation to make the change work successfully.	a. Percent of stakeholder participation in the awareness workshops/sessions b. Level of senior management involvement with I&T risk management programs and strategy c. Percent of users appropriately trained for I&T risk assessment and mitigation d. Amount of market research conducted to measure and support the need for change

I&T Risk-specific Activities (in addition to COBIT 2019 core-model activities)	Capability Level
1. Include I&T risk impact perception in the awareness sessions, so the audience is familiar with the potential threats and opportunities.	2
2. Ensure I&T risk and information security involvement in the change process.	
3. Establish an information risk management culture.	3
4. Assess the impact of the change on the enterprise's risk profile.	
5. Ensure that market research is conducted to identify and measure the need for change to maintain or increase the enterprise's competitive edge.	

Related Guidance (Standards, Frameworks, Compliance Requirements)	Detailed Reference
PROSCI® 3-Phase Change Management Process	Phase 1. Preparing for change—Define your change management strategy

Management Practice	Example I&T Risk-specific Metrics
BAI05.02 Form an effective implementation team. Establish an effective implementation team by assembling appropriate members, creating trust, and establishing common goals and effectiveness measures.	a. Level of participation of I&T risk professionals in the awareness sessions b. Number of qualified I&T risk management professionals in the implementation team

I&T Risk-specific Activities (in addition to COBIT 2019 core-model activities)	Capability Level
1. Include risk professionals on the implementation team.	2
2. Designate qualified I&T risk management professionals to serve on implementation teams.	3
3. Establish a change committee or change board to help ensure proper oversight and approval of deviations from the enterprise's objectives.	

Related Guidance (Standards, Frameworks, Compliance Requirements)	Detailed Reference
PROSCI® 3-Phase Change Management Process	Phase 1. Preparing for change—Prepare your change management team

A. Component: Process *(cont.)*		
Management Practice	**Example I&T Risk-specific Metrics**	
BAI05.03 Communicate desired vision. Communicate the desired vision for the change in the language of those affected by it. The communication should be made by senior management and include the rationale for, and benefits of, the change; the impacts of not making the change; and the vision, the road map and the involvement required of the various stakeholders.	a. Level of participation of I&T risk professionals in the awareness sessions b. Number of communications on the benefits of I&T risk management c. Percent of employees aware of the I&T risk management objectives of the organization d. Level of involvement of the communications team or department	
I&T Risk-specific Activities (in addition to COBIT 2019 core-model activities)		**Capability Level**
1. Communicate the threats and opportunities associated with the desired vision.		2
2. Develop and communicate the vision for the information risk management objectives.		
3. Illustrate and communicate the benefits resulting from I&T risk management.		
4. Communicate changes to the enterprise risk profile (including "positive" risk).		3
Related Guidance (Standards, Frameworks, Compliance Requirements)	**Detailed Reference**	
No related guidance for this management practice		
Management Practice	**Example I&T Risk-specific Metrics**	
BAI05.04 Empower role players and identify short-term wins. Empower those with implementation roles by assigning accountability. Provide training and align organizational structures and HR processes. Identify and communicate short-term wins that are important from a change-enablement perspective.	a. Level of role players' awareness about the underlying I&T risk issues b. Number of employees providing input to I&T risk management c. Percent of roles with defined I&T risk management responsibilities in their job descriptions d. Percent of employees with I&T risk management-related performance metrics	
I&T Risk-specific Activities (in addition to COBIT 2019 core-model activities)		**Capability Level**
1. Embed risk concepts in training while motivating the organizational resources with short-term wins.		2
2. Communicate roles and responsibilities for I&T risk management including assessment, response and escalation to the appropriate staff.		
Related Guidance (Standards, Frameworks, Compliance Requirements)	**Detailed Reference**	
No related guidance for this management practice		
Management Practice	**Example I&T Risk-specific Metrics**	
BAI05.05 Enable operation and use. Plan and implement all technical, operational and usage aspects so all those who are involved in the future state environment can exercise their responsibility.	Additional I&T risk-specific metrics have not been identified for this practice. The COBIT 2019 core metrics are applicable.	
I&T Risk-specific Activities (in addition to COBIT 2019 core-model activities)		**Capability Level**
1. Ensure all risk is managed according to risk tolerance levels.		3
2. Ensure continuous monitoring of KPIs.		
Related Guidance (Standards, Frameworks, Compliance Requirements)	**Detailed Reference**	
PROSCI® 3-Phase Change Management Process	Phase 2. Managing change	

A. Component: Process *(cont.)*		
Management Practice	**Example I&T Risk-specific Metrics**	
BAI05.06 Embed new approaches. Embed new approaches by tracking implemented changes, assessing the effectiveness of the operation and use plan, and sustaining ongoing awareness through regular communication. Take corrective measures as appropriate (which may include enforcing compliance).	Additional I&T risk-specific metrics have not been identified for this practice. The COBIT 2019 core metrics are applicable.	
I&T Risk-specific Activities (in addition to COBIT 2019 core-model activities)		**Capability Level**
1. Ensure that the implemented solution is included in the risk-based audit plan.		2
Related Guidance (Standards, Frameworks, Compliance Requirements)	**Detailed Reference**	
PROSCI® 3-Phase Change Management Process	Phase 3. Reinforcing change	
Management Practice	**Example I&T Risk-specific Metrics**	
BAI05.07 Sustain changes. Sustain changes through effective training of new staff, ongoing communication campaigns, continued commitment of top management, monitoring of adoption and sharing of lessons learned across the enterprise.	Additional I&T risk-specific metrics have not been identified for this practice. The COBIT 2019 core metrics are applicable.	
I&T Risk-specific Activities (in addition to COBIT 2019 core-model activities)		**Capability Level**
1. Ensure continuous risk evaluation.		3
Related Guidance (Standards, Frameworks, Compliance Requirements)	**Detailed Reference**	
PROSCI® 3-Phase Change Management Process	Phase 3. Reinforcing change	

C. Component: Information Flows and Items				
Management Practice	**I&T Risk-specific Inputs**		**I&T Risk-specific Outputs**	
BAI05.01 Establish the desire to change.	**From**	**Description**	**Description**	**To**
	IRFA APO07.03	Risk awareness training curriculum	There are no I&T risk-specific outputs for this practice.	
BAI05.02 Form an effective implementation team.	IRFA APO07.03	Risk awareness training curriculum	There are no I&T risk-specific outputs for this practice.	
BAI05.03 Communicate desired vision.	There are no I&T risk-specific inputs or outputs for this practice.			
BAI05.04 Empower role players and identify short-term wins.	There are no I&T risk-specific inputs or outputs for this practice.			
BAI05.05 Enable operation and use.	There are no I&T risk-specific inputs or outputs for this practice.			
BAI05.06 Embed new approaches.	There are no I&T risk-specific inputs or outputs for this practice.			
BAI05.07 Sustain changes.	There are no I&T risk-specific inputs or outputs for this practice.			
Related Guidance (Standards, Frameworks, Compliance Requirements)	**Detailed Reference**			
No related guidance for this component				

Page intentionally left blank

Domain: Build, Acquire and Implement Management Objective: BAI08 — Managed Knowledge	Focus Area: I&T Risk
Description	
Maintain the availability of relevant, current, validated and reliable knowledge and management information to support all process activities and to facilitate decision making related to the governance and management of enterprise I&T. Plan for the identification, gathering, organizing, maintaining, use and retirement of knowledge.	
Purpose	
Provide the knowledge and information required to support all staff in the governance and management of enterprise I&T and allow for informed decision making.	
I&T Risk Focus Area Relevance	
I&T risk information and knowledge need to be created, developed, acquired, controlled, maintained and secured.	

A. Component: Process		
Management Practice	**Example I&T Risk-specific Metrics**	
BAI08.01 Identify and classify sources of information for governance and management of I&T. Identify, validate and classify diverse sources of internal and external information required to enable governance and management of I&T, including strategy documents, incident reports and configuration information that progresses from development to operations before going live.	a. Number of updates to the I&T risk register to ensure data are current and reliable b. Number of independent sources of information to support the I&T risk management	
I&T Risk-specific Activities (in addition to COBIT 2019 core-model activities)		**Capability Level**
1. Support the use and sharing of information within the risk function, in accordance with the information's classification and sensitivity.		3
2. Develop a structure for how to classify risk management information with the associated risk owners.		
Related Guidance (Standards, Frameworks, Compliance Requirements)	**Detailed Reference**	
No related guidance for this management practice		
Management Practice	**Example I&T Risk-specific Metrics**	
BAI08.02 Organize and contextualize information into knowledge. Organize information based on classification criteria. Identify and create meaningful relationships among information elements and enable use of information. Identify owners, and leverage and implement enterprise-defined information levels of access to management information and knowledge resources.	a. Percent of critical or high I&T risk issues with no mitigation approach	
I&T Risk-specific Activities (in addition to COBIT 2019 core-model activities)		**Capability Level**
1. Map risk management roles to knowledge areas and ensure that proper access control is in place for relevant information.		3
Related Guidance (Standards, Frameworks, Compliance Requirements)	**Detailed Reference**	
COSO Enterprise Risk Management, June 2017	10. Information, Communication, and Reporting - Principle 18	

A. Component: Process *(cont.)*		
Management Practice	**Example I&T Risk-specific Metrics**	
BAI08.03 Use and share knowledge. Propagate available knowledge resources to relevant stakeholders and communicate how these resources can be used to address different needs (e.g., problem solving, learning, strategic planning and decision making).	a. Percent of I&T risk management information mapped to appropriate logical access roles b. Degree to which I&T risk exposure exceeds established I&T risk tolerance thresholds	
I&T Risk-specific Activities (in addition to COBIT 2019 core-model activities)		**Capability Level**
1. Ensure that the proper measures for data risk (loss, theft, corruption, etc.) are in place.		3
2. Implement access controls through the use of policies and processes to restrict unauthorized use and sharing of risk management information.		
Related Guidance (Standards, Frameworks, Compliance Requirements)	**Detailed Reference**	
CMMI Cybermaturity Platform, 2018	PP.IS Apply Information Sharing; IR.ES Ensure Information sharing	
ITIL V3, 2011	Service Transition, 4.7 Knowledge Management	
PMBOK Guide Sixth Edition, 2017	Part 1: 4.4 Manage project knowledge	
Management Practice	**Example I&T Risk-specific Metrics**	
BAI08.04 Evaluate and update or retire information. Measure the use and evaluate the currency and relevance of information. Update information or retire obsolete information.	a. Percent of I&T risk acceptances reviewed at least yearly b. Percent of approved I&T risk acceptances	
I&T Risk-specific Activities (in addition to COBIT 2019 core-model activities)		**Capability Level**
1. Securely dispose of risk management information according to the established and approved data retention policy.		2
2. Send confirmation to the information owner or information custodian about the disposal of risk management information.		
Related Guidance (Standards, Frameworks, Compliance Requirements)	**Detailed Reference**	
No related guidance for this management practice		

C. Component: Information Flows and Items				
Management Practice	**I&T Risk-specific Inputs**		**I&T Risk-specific Outputs**	
	From	**Description**	**Description**	**To**
BAI08.01 Identify and classify sources of information for governance and management of I&T.	There are no I&T risk-specific inputs for this practice.		Classification of risk function information	IRFA BAI08.03
BAI08.02 Organize and contextualize information into knowledge.	There are no I&T risk-specific inputs for this practice.		Published knowledge repositories	IRFA BAI08.03
BAI08.03 Use and share knowledge.	IRFA BAI08.01	Classification of risk function information	Access control over risk management information	Internal
	IRFA BAI08.02	Published knowledge repositories		
BAI08.04 Evaluate and update or retire information.	There are no I&T risk-specific inputs for this practice.		Updated rules for knowledge requirement and information disposal	Internal
Related Guidance (Standards, Frameworks, Compliance Requirements)	**Detailed Reference**			
No related guidance for this component				

4.4 Deliver, Service and Support (DSS)

01 Managed Operations

04 Managed Continuity

06 Managed Business Process Controls

Page intentionally left blank

Domain: Deliver, Service and Support Management Objective: DSS01— Managed Operations	Focus Area: I&T Risk
Description	
Coordinate and execute the activities and operational procedures required to deliver internal and outsourced I&T services. Include the execution of predefined standard operating procedures and the required monitoring activities.	
Purpose	
Deliver I&T operational product and service outcomes as planned.	
I&T Risk Focus Area Relevance	
Provide I&T risk thought leadership to the business to support the assessment and continuous monitoring of operational risk in the enterprise.	

A. Component: Process		
Management Practice	**Example I&T Risk-specific Metrics**	
DSS01.01 Perform operational procedures. Maintain and perform operational procedures and operational tasks reliably and consistently.	a. Percent of operational procedures without IT risk approval	
I&T Risk-specific Activities (in addition to COBIT 2019 core-model activities)		**Capability Level**
1. Ensure co-development of operational procedures to help ensure that I&T risk is considered and minimized.		2
2. Evaluate operational processes to help ensure that completeness and accuracy controls are built in end to end.		
Related Guidance (Standards, Frameworks, Compliance Requirements)	**Detailed Reference**	
CMMI Cybermaturity Platform, 2018	TP.SE Safeguard Operational Environment	
HITRUST CSF version 9, September 2017	09.01 Document Operating Procedures	
ISO/IEC 27002:2013/Cor.2:2015(E)	12.1 Operational procedures and responsibilities	
ITIL V3, 2011	Service Operation, 4.1 Event Management	
National Institute of Standards and Technology Special Publication 800-53, Revision 5 (Draft), August 2017	3.13 Physical and environmental protection (PE-13, PE-14, PE-15)	
Management Practice	**Example I&T Risk-specific Metrics**	
DSS01.02 Manage outsourced I&T services. Manage the operation of outsourced I&T services to maintain the protection of enterprise information and reliability of service delivery.	a. Percent of outsource contracts without right-to-audit clauses	
I&T Risk-specific Activities (in addition to COBIT 2019 core-model activities)		**Capability Level**
1. Ensure that third parties are required to establish and maintain controls that help keep I&T risk levels uniform with those of the enterprise.		3
2. Assist in the periodic evaluation of critical outsource arrangements to help ensure that relevant enterprise I&T risk is within specified limits.		
Related Guidance (Standards, Frameworks, Compliance Requirements)	**Detailed Reference**	
ISF, The Standard of Good Practice for Information Security 2016	SC1.2 Outsourcing	
ISO/IEC 20000-1:2011(E)	4.2 Governance of processes operated by other parties	

A. Component: Process *(cont.)*		
Management Practice	**Example I&T Risk-specific Metrics**	
DSS01.03 Monitor I&T infrastructure. Monitor the I&T infrastructure and related events. Store sufficient chronological information in operations logs to reconstruct and review time sequences of operations and other activities surrounding or supporting operations.	a. Percent of planned vs. unplanned downtime b. Percent of on-premise vs. off-premise computing environments	
I&T Risk-specific Activities (in addition to COBIT 2019 core-model activities)		**Capability Level**
1. Deploy log aggregation mechanisms across the network.		2
2. Develop a risk-based approach to identify and respond to I&T-related threats and events.		3
Related Guidance (Standards, Frameworks, Compliance Requirements)	**Detailed Reference**	
National Institute of Standards and Technology Special Publication 800-53, Revision 5 (Draft), August 2017	3.10 Maintenance (MA-2, MA-3)	
Management Practice	**Example I&T Risk-specific Metrics**	
DSS01.04 Manage the environment. Maintain measures for protection against environmental factors. Install specialized equipment and devices to monitor and control the environment.	a. Percent of devices without data loss prevention controls	
I&T Risk-specific Activities (in addition to COBIT 2019 core-model activities)		**Capability Level**
1. Establish monitoring mechanisms, across all devices, to help reduce the risk of intentional or accidental data loss.		3
Related Guidance (Standards, Frameworks, Compliance Requirements)	**Detailed Reference**	
National Institute of Standards and Technology Special Publication 800-37, Revision 2 (Draft), May 2018	2.1 System and system elements; 3.2 Categorization (Task 5, 6)	
Management Practice	**Example I&T Risk-specific Metrics**	
DSS01.05 Manage facilities. Manage facilities, including power and communications equipment, in line with laws and regulations, technical and business requirements, vendor specifications, and health and safety guidelines.	a. Percent of planned vs. unplanned maintenance	
I&T Risk-specific Activities (in addition to COBIT 2019 core-model activities)		**Capability Level**
1. Ensure that security and resilience mechanisms are in place to reduce risk of critical infrastructure intrusion or attack.		3
2. Establish information-sharing channels between internal and external stakeholders responsible for development of risk mitigation strategies.		
Related Guidance (Standards, Frameworks, Compliance Requirements)	**Detailed Reference**	
National Institute of Standards and Technology Special Publication 800-37, Revision 2, September 2017	3.2 Categorization (Task 5, 6): Inputs and Outputs	

C. Component: Information Flows and Items				
Management Practice	**I&T Risk-specific Inputs**		**I&T Risk-specific Outputs**	
IRFA DSS01.01 Perform operational procedures.	**From**	**Description**	**Description**	**To**
	There are no I&T risk-specific inputs or outputs for this practice.			
IRFA DSS01.02 Manage outsourced I&T services.	IRFA APO10.01	Vendor risk management plan	There are no I&T risk-specific outputs for this practice.	
IRFA DSS01.03 Monitor I&T infrastructure.	IRFA MEA01.01	Risk management monitoring process and procedure	Risk identified from new threats or events	IRFA DSS06.06
IRFA DSS01.04 Manage the environment.	IRFA MEA01.01	Risk management monitoring process and procedure	There are no I&T risk-specific outputs for this practice.	
IRFA DSS01.05 Manage facilities.	IRFA MEA01.01	Risk management monitoring process and procedure	There are no I&T risk-specific outputs for this practice.	

Related Guidance (Standards, Frameworks, Compliance Requirements)	Detailed Reference
National Institute of Standards and Technology Special Publication 800-37, Revision 2, September 2017	3.2 Categorization (Task 5, 6): Inputs and Outputs

Page intentionally left blank

Domain: Deliver, Service and Support Management Objective: DSS04—Managed Continuity	Focus Area: I&T Risk
Description	
Establish and maintain a plan to enable the business and IT organizations to respond to incidents and quickly adapt to disruptions. This will enable continued operations of critical business processes and required I&T services and maintain availability of resources, assets and information at a level acceptable to the enterprise.	
Purpose	
Adapt rapidly, continue business operations and maintain availability of resources and information at a level acceptable to the enterprise in the event of a significant disruption (e.g., threats, opportunities, demands).	
I&T Risk Focus Area Relevance	
Strategies must be promoted to ensure key I&T risk scenarios (e.g., geopolitical, information, infrastructure, acts of nature) are incorporated into the enterprise resiliency program.	

A. Component: Process		
Management Practice	**Example I&T Risk-specific Metrics**	
DSS04.01 Define the business continuity policy, objectives and scope. Define business continuity policy and scope, aligned with enterprise and stakeholder objectives, to improve business resilience.	Additional I&T risk-specific metrics have not been identified for this practice. The COBIT 2019 core metrics are applicable.	
I&T Risk-specific Activities (in addition to COBIT 2019 core-model activities)		**Capability Level**
Additional I&T risk-specific activities have not been identified for this practice. The COBIT 2019 core activities are applicable.		N/A
Related Guidance (Standards, Frameworks, Compliance Requirements)	**Detailed Reference**	
HITRUST CSF version 9, September 2017	12.01 Information Security Aspects of Business Continuity Management	
ISF, The Standard of Good Practice for Information Security 2016	BC1.1 Business Continuity Strategy; BC1.2 Business Continuity Programme	
ISO/IEC 27002:2013/Cor.2:2015(E)	17. Information security aspects of business continuity management	
National Institute of Standards and Technology Special Publication 800-53, Revision 5 (Draft), August 2017	3.6 Contingency planning (CP-1)	
Management Practice	**Example I&T Risk-specific Metrics**	
DSS04.02 Maintain business resilience. Evaluate business resilience options and choose a cost-effective and viable strategy that will ensure enterprise continuity, disaster recovery and incident response in the face of a disaster or other major incident or disruption.	a. Percent of outdated business continuity and/or disaster recovery plans b. Percent of BCP and DRP plans not approved by key stakeholders	
I&T Risk-specific Activities (in addition to COBIT 2019 core-model activities)		**Capability Level**
1. Determine who is responsible for maintaining the plan, and review the maintenance and revision procedures.		2
2. Encourage I&T management to define and execute BCP and DRP change control procedures to help ensure that they are kept up to date and reflect actual business requirements.		
3. Ensure that changes to the BCP and DRP procedures and responsibilities are communicated clearly and in a timely manner.		3
Related Guidance (Standards, Frameworks, Compliance Requirements)	**Detailed Reference**	
ISF, The Standard of Good Practice for Information Security 2016	BC1.3 Resilient Technical Environments	
ITIL V3, 2011	Service Design, 4.6 IT Continuity Management	
National Institute of Standards and Technology Special Publication 800-53, Revision 5 (Draft), August 2017	3.6 Contingency planning (CP-2)	

A. Component: Process *(cont.)*		
Management Practice	**Example I&T Risk-specific Metrics**	
DSS04.03 Develop and implement a business continuity response. Develop a business continuity plan (BCP) and disaster recovery plan (DRP) based on the strategy. Document all procedures necessary for the enterprise to continue critical activities in the event of an incident.	a. Percent of continuity plans for which a risk assessment was not performed	
I&T Risk-specific Activities (in addition to COBIT 2019 core-model activities)		**Capability Level**
1. Provide all concerned parties with regular training sessions regarding the BCP and DRP procedures and their roles and responsibilities in case of an incident or disaster.		2
2. Verify and enhance training according to the results of the contingency tests.		
3. Determine if a risk assessment has been performed that identifies business and operational exposures.		3
4. Ensure that damage assessment steps with formal decision points and thresholds are in place to activate the plan.		
Related Guidance (Standards, Frameworks, Compliance Requirements)	**Detailed Reference**	
ISF, The Standard of Good Practice for Information Security 2016	BC1.4 Crisis Management; BC2.1 Business Continuity Planning	
National Institute of Standards and Technology Special Publication 800-53, Revision 5 (Draft), August 2017	3.6 Contingency planning (CP-6, CP-9, CP-10)	
Management Practice	**Example I&T Risk-specific Metrics**	
DSS04.04 Exercise, test and review the business continuity plan (BCP) and disaster response plan (DRP). Test continuity on a regular basis to exercise plans against predetermined outcomes, uphold business resilience and allow innovative solutions to be developed.	a. Percent of BCP and/or DRP issues identified previously that have not been remediated	
I&T Risk-specific Activities (in addition to COBIT 2019 core-model activities)		**Capability Level**
1. Confirm that I&T continuity plans are tested on a regular basis to ensure that I&T systems can be effectively recovered, shortcomings are addressed and the plan remains relevant.		2
2. Ensure the preparation, documentation and reporting of test results. If called for by the test results, ensure that an action plan is implemented.		
3. Ensure that the testing plan incorporates recovery of single applications, integrated testing scenarios, end-to-end testing and integrated vendor testing.		3
4. Plan the actions to be taken for the period when I&T is recovering and resuming services (e.g., activate backup site, initiate alternative processing, communicate to customers).		
5. Ensure that the business understands I&T recovery times and the necessary technology investments to support business recovery and resumption needs.		
Related Guidance (Standards, Frameworks, Compliance Requirements)	**Detailed Reference**	
CMMI Cybermaturity Platform, 2018	PP.RS Develop and Maintain Response Plans; PP.RP Develop and Maintain Recovery Plans	
ISF, The Standard of Good Practice for Information Security 2016	BC2.3 Business Continuity Testing	
The CIS Critical Security Controls for Effective Cyber Defense Version 6.1, August 2016	CSC 20: Penetration Tests and Red Team Exercises	

A. Component: Process *(cont.)*	
Management Practice	**Example I&T Risk-specific Metrics**
DSS04.05 Review, maintain and improve the continuity plans. Conduct a management review of the continuity capability at regular intervals to ensure its continued suitability, adequacy and effectiveness. Manage changes to the plans in accordance with the change control process to ensure that continuity plans are kept up to date and continually reflect actual business requirements.	Additional I&T risk-specific metrics have not been identified for this practice. The COBIT 2019 core metrics are applicable.
I&T Risk-specific Activities (in addition to COBIT 2019 core-model activities)	**Capability Level**
Additional I&T risk-specific activities have not been identified for this practice. The COBIT 2019 core activities are applicable.	N/A
Related Guidance (Standards, Frameworks, Compliance Requirements)	**Detailed Reference**
No related guidance for this management practice	
Management Practice	**Example I&T Risk-specific Metrics**
DSS04.06 Conduct continuity plan training. Provide all concerned internal and external parties with regular training sessions regarding procedures and their roles and responsibilities in case of disruption.	Additional I&T risk-specific metrics have not been identified for this practice. The COBIT 2019 core metrics are applicable.
I&T Risk-specific Activities (in addition to COBIT 2019 core-model activities)	**Capability Level**
Additional I&T risk-specific activities have not been identified for this practice. The COBIT 2019 core activities are applicable.	N/A
Related Guidance (Standards, Frameworks, Compliance Requirements)	**Detailed Reference**
National Institute of Standards and Technology Special Publication 800-53, Revision 5 (Draft), August 2017	3.6 Contingency planning (CP-4)
Management Practice	**Example I&T Risk-specific Metrics**
DSS04.07 Manage backup arrangements. Maintain availability of business-critical information.	Additional I&T risk-specific metrics have not been identified for this practice. The COBIT 2019 core metrics are applicable.
I&T Risk-specific Activities (in addition to COBIT 2019 core-model activities)	**Capability Level**
Additional I&T risk-specific activities have not been identified for this practice. The COBIT 2019 core activities are applicable.	N/A
Related Guidance (Standards, Frameworks, Compliance Requirements)	**Detailed Reference**
CMMI Cybermaturity Platform, 2018	IP.BP Apply Backup Processes
HITRUST CSF version 9, September 2017	09.05 Information Back-Up
ISF, The Standard of Good Practice for Information Security 2016	SY2.3 Backup
ISO/IEC 27002:2013/Cor.2:2015(E)	12.3 Backup
National Institute of Standards and Technology Special Publication 800-53, Revision 5 (Draft), August 2017	3.6 Contingency planning (CP-3)
The CIS Critical Security Controls for Effective Cyber Defense Version 6.1, August 2016	CSC 10: Data Recovery Capability

A. Component: Process *(cont.)*		
Management Practice	**Example I&T Risk-specific Metrics**	
DSS04.08 Conduct post-resumption review. Assess the adequacy of the business continuity plan (BCP) and disaster response plan (DRP) following successful resumption of business processes and services after a disruption.	Additional I&T risk-specific metrics have not been identified for this practice. The COBIT 2019 core metrics are applicable.	
I&T Risk-specific Activities (in addition to COBIT 2019 core-model activities)		**Capability Level**
Additional I&T risk-specific activities have not been identified for this practice. The COBIT 2019 core activities are applicable.		N/A
Related Guidance (Standards, Frameworks, Compliance Requirements)	**Detailed Reference**	
No related guidance for this management practice		

C. Component: Information Flows and Items				
Management Practice	**I&T Risk-specific Inputs**		**I&T Risk-specific Outputs**	
	From	**Description**	**Description**	**To**
DSS04.01 Define the business continuity policy, objectives and scope.	There are no I&T risk-specific inputs or outputs for this practice.			
DSS04.02 Maintain business resilience.	IRFA APO02.06	Risk management plan	There are no I&T risk-specific outputs for this practice.	
DSS04.03 Develop and implement a business continuity response.	There are no I&T risk-specific inputs or outputs for this practice.			
DSS04.04 Exercise, test and review the business continuity plan (BCP) and disaster response plan (DRP).	There are no I&T risk-specific inputs for this practice.		Newly identified risk in BCP/DRP	IRFA DSS04.05
DSS04.05 Review, maintain and improve the continuity plans.	IRFA DSS04.04	Newly identified risk in BCP/DRP	There are no I&T risk-specific outputs for this practice.	
DSS04.06 Conduct continuity plan training.	There are no I&T risk-specific inputs or outputs for this practice.			
DSS04.07 Manage backup arrangements.	There are no I&T risk-specific inputs or outputs for this practice.			
DSS04.08 Conduct post-resumption review.	Internal	Risk management plan	There are no I&T risk-specific outputs for this practice.	
Related Guidance (Standards, Frameworks, Compliance Requirements)	**Detailed Reference**			
No related guidance for this component				

Domain: Deliver, Service and Support Management Objective: DSS06—Managed Business Process Controls	Focus Area: I&T Risk
Description	
Define and maintain appropriate business process controls to ensure that information related to and processed by in-house or outsourced business processes satisfies all relevant information control requirements. Identify the relevant information control requirements. Manage and operate adequate input, throughput and output controls (application controls) to ensure that information and information processing satisfy these requirements.	
Purpose	
Maintain information integrity and the security of information assets handled within business processes in the enterprise or its outsourced operation.	
I&T Risk Focus Area Relevance	
Ensure business processes are controlled by policies, procedures and structures that create value to the enterprise and provide reasonable assurance that a business process will achieve its objectives while mitigating risk.	

A. Component: Process		
Management Practice	**Example I&T Risk-specific Metrics**	
DSS06.01 Align control activities embedded in business processes with enterprise objectives. Continually assess and monitor the execution of business process activities and related controls (based on enterprise risk), to ensure that processing controls align with business needs.	Additional I&T risk-specific metrics have not been identified for this practice. The COBIT 2019 core metrics are applicable.	
I&T Risk-specific Activities (in addition to COBIT 2019 core-model activities)		**Capability Level**
Additional I&T risk-specific activities have not been identified for this practice. The COBIT 2019 core activities are applicable.		N/A
Related Guidance (Standards, Frameworks, Compliance Requirements)	**Detailed Reference**	
National Institute of Standards and Technology Special Publication 800-37, Revision 2 (Draft), May 2018	3.1 Preparation (Task 10, 11)	
The CIS Critical Security Controls for Effective Cyber Defense Version 6.1, August 2016	CSC 14: Controlled Access Based on the Need to Know	
Management Practice	**Example I&T Risk-specific Metrics**	
DSS06.02 Control the processing of information. Operate the execution of the business process activities and related controls, based on enterprise risk. Ensure that information processing is valid, complete, accurate, timely and secure (i.e., reflects legitimate and authorized business use).	Additional I&T risk-specific metrics have not been identified for this practice. The COBIT 2019 core metrics are applicable.	
I&T Risk-specific Activities (in addition to COBIT 2019 core-model activities)		**Capability Level**
Additional I&T risk-specific activities have not been identified for this practice. The COBIT 2019 core activities are applicable.		N/A
Related Guidance (Standards, Frameworks, Compliance Requirements)	**Detailed Reference**	
HITRUST CSF version 9, September 2017	13.01 Openness and Transparency; 13.02 Individual Choice and Participation	
ISF, The Standard of Good Practice for Information Security 2016	BA1.4 Information Validation	

A. Component: Process *(cont.)*		
Management Practice	**Example I&T Risk-specific Metrics**	
DSS06.03 Manage roles, responsibilities, access privileges and levels of authority. Manage business roles, responsibilities, levels of authority and segregation of duties needed to support the business process objectives. Authorize access to all information assets related to business information processes, including those under the custody of the business, IT and third parties. This ensures that the business knows where the data are and who is handling data on its behalf.	Additional I&T risk-specific metrics have not been identified for this practice. The COBIT 2019 core metrics are applicable.	
I&T Risk-specific Activities (in addition to COBIT 2019 core-model activities)		**Capability Level**
Additional I&T risk-specific activities have not been identified for this practice. The COBIT 2019 core activities are applicable.		N/A
Related Guidance (Standards, Frameworks, Compliance Requirements)	**Detailed Reference**	
CMMI Cybermaturity Platform, 2018	AC.MP Manage Access Permissions	
The CIS Critical Security Controls for Effective Cyber Defense Version 6.1, August 2016	CSC 18: Application Software Security	
Management Practice	**Example I&T Risk-specific Metrics**	
DSS06.04 Manage errors and exceptions. Manage business process exceptions and errors and facilitate remediation, executing defined corrective actions and escalating as necessary. This treatment of exceptions and errors provides assurance of the accuracy and integrity of the business information process.	a. Percent of unresolved automated business process errors	
I&T Risk-specific Activities (in addition to COBIT 2019 core-model activities)		**Capability Level**
1. Ensure that procedures are developed to identify and resolve the risk of errors in automated business processes.		3
Related Guidance (Standards, Frameworks, Compliance Requirements)	**Detailed Reference**	
No related guidance for this management practice		
Management Practice	**Example I&T Risk-specific Metrics**	
DSS06.05 Ensure traceability and accountability for information events. Ensure that business information can be traced to an originating business event and associated with accountable parties. This discoverability provides assurance that business information is reliable and has been processed in accordance with defined objectives.	a. Percent of key manual business processes	
I&T Risk-specific Activities (in addition to COBIT 2019 core-model activities)		**Capability Level**
1. Work with the business to identify ways to streamline business processes and monitor controls via business process automation.		3
2. Review the adequacy of the documentation and mapping of manual business processes and defined automation technology requirements.		
Related Guidance (Standards, Frameworks, Compliance Requirements)	**Detailed Reference**	
No related guidance for this management practice		

A. Component: Process *(cont.)*		
Management Practice	**Example I&T Risk-specific Metrics**	
DSS06.06 Secure information assets. Secure information assets accessible by the business through approved methods, including information in electronic form (e.g., portable media devices, user applications and storage devices, or other methods that create new assets in any form), information in physical form (e.g., source documents or output reports) and information during transit. This benefits the business by providing end-to-end safeguarding of information.	Additional I&T risk-specific metrics have not been identified for this practice. The COBIT 2019 core metrics are applicable.	
I&T Risk-specific Activities (in addition to COBIT 2019 core-model activities)		**Capability Level**
Additional I&T risk-specific activities have not been identified for this practice. The COBIT 2019 core activities are applicable.		N/A
Related Guidance (Standards, Frameworks, Compliance Requirements)	**Detailed Reference**	
CMMI Cybermaturity Platform, 2018	AC.MP Manage Access Permissions	
The CIS Critical Security Controls for Effective Cyber Defense Version 6.1, August 2016	CSC 18: Application Software Security	
National Institute of Standards and Technology Special Publication 800-37, Revision 2, September 2017	3.1 Preparation (Task 10, 11): Inputs and Outputs	

C. Component: Information Flows and Items				
Management Practice	**I&T Risk-specific Inputs**		**I&T Risk-specific Outputs**	
	From	**Description**	**Description**	**To**
DSS06.01 Align control activities embedded in business processes with enterprise objectives.	There are no I&T risk-specific inputs or outputs for this practice.			
DSS06.02 Control the processing of information.	IRFA APO14.04	Risk data quality management policy	There are no I&T risk-specific outputs for this practice.	
DSS06.03 Manage roles, responsibilities, access privileges and levels of authority.	There are no I&T risk-specific inputs or outputs for this practice.			
DSS06.04 Manage errors and exceptions.	IRFA BAI03.09	Record of risk associated with approved changes	Issues log	IRFA MEA02.04
DSS06.05 Ensure traceability and accountability for information events.	There are no I&T risk-specific inputs for this practice.		System generated audit log	Internal
DSS06.06 Secure information assets.	IRFA DSS01.03	Risk identified from new threats or events	There are no I&T risk-specific outputs for this practice.	
Related Guidance (Standards, Frameworks, Compliance Requirements)	**Detailed Reference**			
National Institute of Standards and Technology Special Publication 800-37, Revision 2, September 2017	3.1 Preparation (Task 10, 11): Inputs and Outputs			

Page intentionally left blank

4.5 Monitor, Evaluate and Assess (MEA)

Page intentionally left blank

Domain: Monitor, Evaluate and Assess Management Objective: MEA01 — Managed Performance and Conformance Monitoring	Focus Area: I&T Risk
Description	
Collect, validate and evaluate enterprise and alignment goals and metrics. Monitor that processes and practices are performing against agreed performance and conformance goals and metrics. Provide reporting that is systematic and timely.	
Purpose	
Provide transparency of performance and conformance and drive achievement of goals.	
I&T Risk Focus Area Relevance	
Monitoring, evaluating and assessing I&T risk is a key aspect of managing effective business and IT.	

A. Component: Process

Management Practice	Example I&T Risk-specific Metrics
MEA01.01 Establish a monitoring approach. Engage with stakeholders to establish and maintain a monitoring approach to define the objectives, scope and method for measuring business solution and service delivery and contribution to enterprise objectives. Integrate this approach with the corporate performance management system.	a. Percent of I&T approved risk issues with defined metrics b. Percent of I&T risk metrics that are periodically reviewed c. Percent of I&T risk metrics that are aligned to business risk issues d. Percent of I&T risk issues with new business strategy changes

I&T Risk-specific Activities (in addition to COBIT 2019 core-model activities)	Capability Level
1. Identify and confirm risk management stakeholders.	2
2. Engage with stakeholders and communicate the risk management requirements and objectives for monitoring and reporting.	
3. Align and continually maintain the risk management monitoring and evaluation approach with the I&T and enterprise approaches.	3
4. Request, prioritize and allocate resources for monitoring risk management.	
5. Define responsibilities in terms of risk management monitoring and reporting.	
6. Establish risk-based monitoring and reporting techniques.	
7. Establish the risk management monitoring process and procedure.	4
8. Agree on a life cycle management and change control process for risk management monitoring and reporting.	

Related Guidance (Standards, Frameworks, Compliance Requirements)	Detailed Reference
CMMI Data Management Maturity Model, 2014	Supporting Processes - Measurement and Analysis
SF, The Standard of Good Practice for Information Security 2016	SI2 Security Performance
ISO/IEC 27001:2013/Cor.2:2015(E)	9.1 Monitoring, measurement, analysis and evaluation
ISO/IEC 27004:2016(E)	6. Characteristics; 7. Types of measures; 8. Processes
ISO/IEC 38500:2015(E)	5.5 Principle 4: Performance; 5.6 Principle 5: Conformance
National Institute of Standards and Technology Special Publication 800-37, Revision 2 (Draft), May 2018	3.1 Preparation (Task 13); 3.3 Selection (Task 2); 3.7 Monitoring (Task 1)
National Institute of Standards and Technology Special Publication 800-53, Revision 5 (Draft), August 2017	3.4 Assessment, authorization and monitoring (CA-2, CA-7); 3.20 System and information integrity (SI-4)

A. Component: Process *(cont.)*	
Management Practice	**Example I&T Risk-specific Metrics**
MEA01.02 Set performance and conformance targets. Work with stakeholders to define, periodically review, update and approve performance and conformance targets within the performance measurement system.	a. Percent of I&T risk metrics approved by stakeholders b. Percent of internal controls monitored to ensure I&T risk is mitigated or contained c. Percent of I&T processes that meet defined I&T risk management requirements d. Percent of achieved KPIs

I&T Risk-specific Activities (in addition to COBIT 2019 core-model activities)	Capability Level
1. Define risk management performance targets consistent with overall I&T performance standards.	2
2. Communicate risk management performance and conformance targets with key due diligence stakeholders.	
3. Evaluate whether the risk management goals and metrics are adequate (i.e., specific, measurable, achievable, relevant and time-bound).	

Related Guidance (Standards, Frameworks, Compliance Requirements)	Detailed Reference
CMMI Data Management Maturity Model, 2014	Supporting Processes - Process Management
National Institute of Standards and Technology Special Publication 800-53, Revision 5 (Draft), August 2017	3.4 Assessment, authorization and monitoring (CA-5)

Management Practice	Example I&T Risk-specific Metrics
MEA01.03 Collect and process performance and conformance data. Collect and process timely and accurate data aligned with enterprise approaches.	a. Percent of compliance risk issues (e.g., laws, regulations and contractual requirements) not monitored, evaluated or assessed b. Percent of critical and high I&T risk processes monitored c. Percent of I&T risk management practices that satisfy internal compliance requirements d. Percent of critical and high I&T risk processes not monitored e. Number of key people who can be affected by critical and high-risk situations

I&T Risk-specific Activities (in addition to COBIT 2019 core-model activities)	Capability Level
1. Collect and analyze performance and conformance data relating to risk management.	2
2. Investigate and deploy a governance, risk and compliance (GRC) tool.	3

Related Guidance (Standards, Frameworks, Compliance Requirements)	Detailed Reference
National Institute of Standards and Technology Special Publication 800-53, Revision 5 (Draft), August 2017	3.20 System and information integrity (SI-2)

Management Practice	Example I&T Risk-specific Metrics
MEA01.04 Analyze and report performance. Periodically review and report performance against targets. Use a method that provides a succinct all-around view of I&T performance and fits within the enterprise monitoring system.	a. Percent of identified I&T risk issues for which controls have been tested and assessed as effectively managing the I&T risk within the organization's I&T risk appetite b. Percent of unidentified risk issues resulting in I&T issues c. Percent of survey results measuring satisfaction with the delivery of I&T risk function services

I&T Risk-specific Activities (in addition to COBIT 2019 core-model activities)	Capability Level
1. Design, implement and agree on a range of risk management performance reports.	3
2. Identify areas of nonconformance to compliance and regulations.	
3. Compare the performance values to internal targets and benchmarks.	4

Related Guidance (Standards, Frameworks, Compliance Requirements)	Detailed Reference
CMMI Data Management Maturity Model, 2014	Supporting Processes - Measurement and Analysis
National Institute of Standards and Technology Special Publication 800-53, Revision 5 (Draft), August 2017	3.3 Audit and accountability (AU-6)

A. Component: Process *(cont.)*	
Management Practice	**Example I&T Risk-specific Metrics**
MEA01.05 Ensure the implementation of corrective actions. Assist stakeholders in identifying, initiating and tracking corrective actions to address anomalies.	a. Number of I&T risk deficiencies with action plans identified b. Percent of I&T risk remediation plans not closed timely c. Number of deficiencies effectively remediated by action plans d. Number of deficiencies without action plans or with open-ended target dates

I&T Risk-specific Activities (in addition to COBIT 2019 core-model activities)	Capability Level
1. Identify roles and responsibilities for corrective actions.	3
2. Track that corrective actions have been taken and ensure that they are defined to testify to and sustain the implementation of corrective actions.	
3. Develop and implement a mitigation process based on criticality and ensure that accountability is well defined.	
4. Investigate and deploy a recommendation tracking tool.	
5. Make sure that adequate resources are defined and allocated to implement corrective actions.	

Related Guidance (Standards, Frameworks, Compliance Requirements)	Detailed Reference
ITIL V3, 2011	Continual Service Improvement, 4.1 The 7-Step Improvement Process
National Institute of Standards and Technology Special Publication 800-37, Revision 2 (Draft), May 2018	3.7 Monitoring (Task 3)
National Institute of Standards and Technology Special Publication 800-53, Revision 5 (Draft), August 2017	3.3 Audit and accountability (AU-5)

C. Component: Information Flows and Items				
Management Practice	**I&T Risk-specific Inputs**		**I&T Risk-specific Outputs**	
	From	**Description**	**Description**	**To**
MEA01.01 Establish a monitoring approach.	IRFA EDM01.01	Stakeholder requirements specific to risk priorities and objectives	There are no I&T risk-specific outputs for this practice.	
	IRFA APO11.01	Relevant risk function best practices and standards	Risk management monitoring process and procedure	IRFA DSS01.03 IRFA DSS01.04 IRFA DSS01.05 IRFA MEA01.02
	IRFA APO11.04	Risk function quality metrics implemented in line with best practices		
MEA01.02 Set performance and conformance targets.	IRFA MEA01.01	Risk management monitoring process and procedure	Agreed-on risk management metrics and targets	IRFA MEA01.04
MEA01.03 Collect and process performance and conformance data.	Outside COBIT	Applicable regulations	Processed monitoring data	Internal
MEA01.04 Analyze and report performance.	IRFA MEA01.02	Agreed-on risk management metrics and targets	Risk management performance reports and corrective action plan update	IRFA MEA04.05
MEA01.05 Ensure the implementation of corrective actions.	IRFA MEA01.04	Risk management performance reports and corrective action plan update	Tracking process for corrective actions on risk management issues	Internal
	Outside COBIT	Escalation guidelines		

Related Guidance (Standards, Frameworks, Compliance Requirements)	Detailed Reference
National Institute of Standards and Technology Special Publication 800-37, Revision 2, September 2017	3.1 Preparation (Task 13): Inputs and Outputs; 3.3 Selection (Task 2): Inputs and Outputs; 3.7 Monitoring (Task 1, Task 3): Inputs and Outputs

Domain: Monitor, Evaluate and Assess Management Objective: MEA02 — Managed System of Internal Control	Focus Area: I&T Risk
Description	
Continuously monitor and evaluate the control environment, including self-assessments and self-awareness. Enable management to identify control deficiencies and inefficiencies and to initiate improvement actions. Plan, organize and maintain standards for internal control assessment and process control effectiveness.	
Purpose	
Obtain transparency for key stakeholders on the adequacy of the system of internal controls and thus provide trust in operations, confidence in the achievement of enterprise objectives and an adequate understanding of residual risk.	
I&T Risk Focus Area Relevance	
Managing, monitoring and evaluating the I&T control environment are key activities for I&T risk management.	

A. Component: Process	
Management Practice	**Example I&T Risk-specific Metrics**
MEA02.01 Monitor internal controls. Continuously monitor, benchmark and improve the I&T control environment and control framework to meet organizational objectives.	a. Percent of processes that satisfy I&T risk management control requirements b. Percent of I&T risk issues for which the controls that manage the I&T risk issues are subject to assurance assessment c. Percent of controls that have an assigned control owner d. Frequency of monitoring controls e. Percent of failed controls

I&T Risk-specific Activities (in addition to COBIT 2019 core-model activities)	Capability Level
1. Monitor I&T risk action plans and assess the performance of key I&T risk management practices and whether I&T risk is being managed in line with risk appetite and tolerance.	3
2. Provide independent assurance over I&T risk management.	
3. Ensure that control activities are in place and exceptions are promptly reported, followed up and analyzed.	
4. Ensure that appropriate corrective actions are prioritized and implemented according to the risk management profile (e.g., classify certain exceptions as key risks and others as non-key risks).	
5. Maintain the internal control system, considering ongoing changes in business and I&T risk, processes and the organizational control environment. If gaps exist, evaluate and recommend changes.	

Related Guidance (Standards, Frameworks, Compliance Requirements)	Detailed Reference
HITRUST CSF version 9, September 2017	09.10 Monitoring
ISO/IEC 38502:2017(E)	5.5 Governance and internal control
National Institute of Standards and Technology Special Publication 800-53, Revision 5 (Draft), August 2017	3.3 Audit and accountability (AU-2)

A. Component: Process *(cont.)*	
Management Practice	**Example I&T Risk-specific Metrics**
MEA02.02 Review effectiveness of business process controls. Review the operation of controls, including monitoring and test evidence, to ensure that controls within business processes operate effectively. Include activities to maintain evidence of the effective operation of controls through mechanisms such as periodic testing, continuous monitoring, independent assessments, command and control centers, and network operation centers. This evidence assures the enterprise that controls meet requirements related to business, regulatory and social responsibilities.	a. Percent of controls meeting I&T risk management control requirements b. Percent of control self-assessments rated with no substantiating evidence c. Percent of controls performed by second (e.g., I&T risk management, quality or financial control) and third (i.e., audit) line of defense

I&T Risk-specific Activities (in addition to COBIT 2019 core-model activities)	Capability Level
1. Measure the effectiveness of management controls to mitigate risk.	4
2. Leverage the reviews of the second and third lines of defense to measure the effectiveness of management controls to mitigate I&T risk.	

Related Guidance (Standards, Frameworks, Compliance Requirements)	Detailed Reference
No related guidance for this management practice	

Management Practice	Example I&T Risk-specific Metrics
MEA02.03 Perform control self-assessments. Encourage management and process owners to improve controls proactively through a continuing program of self-assessment that evaluates the completeness and effectiveness of management's control over processes, policies and contracts.	a. Percent of assurance (internal or external) issues not identified in the control self-assessment

I&T Risk-specific Activities (in addition to COBIT 2019 core-model activities)	Capability Level
1. Perform risk management assessments (independent and self-assessment) to identify control adequacy and effectiveness.	3

Related Guidance (Standards, Frameworks, Compliance Requirements)	Detailed Reference
ISO/IEC 27001:2013/Cor.2:2015(E)	9.3 Management review
National Institute of Standards and Technology Special Publication 800-37, Revision 2 (Draft), May 2018	3.7 Monitoring (Task 2)

Management Practice	Example I&T Risk-specific Metrics
MEA02.04 Identify and report control deficiencies. Identify control deficiencies and analyze and identify their underlying root causes. Escalate control deficiencies and report to stakeholders.	a. Percent of control I&T risk assessments performed within agreed timelines b. Percent of control deficiencies

I&T Risk-specific Activities (in addition to COBIT 2019 core-model activities)	Capability Level
1. Review realized risk issues (e.g., incidents) to evaluate the effectiveness of the enterprise's risk management program.	3
2. Review audit reports to identify weaknesses and corrective actions.	

Related Guidance (Standards, Frameworks, Compliance Requirements)	Detailed Reference
No related guidance for this management practice	

C. Component: Information Flows and Items				
Management Practice	**I&T Risk-specific Inputs**		**I&T Risk-specific Outputs**	
MEA02.01 Monitor internal controls.	Outside COBIT	Independent external audits	Defined risk management assurance scope and approach to assess internal controls	IRFA MEA02.03
MEA02.02 Review effectiveness of business process controls.	There are no I&T risk-specific inputs for this practice.		Evidence of effectiveness of risk management	Internal
MEA02.03 Perform control self-assessments.	IRFA MEA02.01	Defined risk management assurance scope and approach to assess internal controls	Risk management assurance assessments	IRFA MEA02.04
MEA02.04 Identify and report control deficiencies.	IRFA MEA02.03	Risk management assurance assessments	Assessment results and remedial actions	IRFA MEA04.08
	IRFA DSS06.04	Issues log		

Related Guidance (Standards, Frameworks, Compliance Requirements)	Detailed Reference
National Institute of Standards and Technology Special Publication 800-37, Revision 2, September 2017	3.7 Monitoring (Task 2): Inputs and Outputs

Page intentionally left blank

Domain: Monitor, Evaluate and Assess Management Objective: MEA03 — Managed Compliance With External Requirements	Focus Area: I&T Risk
Description	
Evaluate that I&T processes and I&T-supported business processes are compliant with laws, regulations and contractual requirements. Obtain assurance that the requirements have been identified and complied with; integrate IT compliance with overall enterprise compliance.	
Purpose	
Ensure that the enterprise is compliant with all applicable external requirements.	
I&T Risk Focus Area Relevance	
Compliance with laws, regulations and contractual requirements is an integral part of I&T risk management.	

A. Component: Process	
Management Practice	**Example I&T Risk-specific Metrics**
MEA03.01 Identify external compliance requirements. On a continuous basis, monitor changes in local and international laws, regulations and other external requirements and identify mandates for compliance from an I&T perspective.	a. Percent of regulatory obligations reflected in the I&T risk register b. Frequency of compliance reviews

I&T Risk-specific Activities (in addition to COBIT 2019 core-model activities)	Capability Level
1. Identify risk management compliance targets for external requirements.	2
2. Determine external compliance requirements to be met (including legal, regulatory, privacy and contractual).	
3. Monitor compliance requirements and determine the consequences of noncompliance.	
4. Communicate risk treatment to appropriate stakeholders.	3

Related Guidance (Standards, Frameworks, Compliance Requirements)	Detailed Reference
CMMI Cybermaturity Platform, 2018	BC.RR Determine Legal/Regulatory Requirements
HITRUST CSF version 9, September 2017	06.01 Compliance with Legal Requirements
ISF, The Standard of Good Practice for Information Security 2016	SM2.3 Legal and Regulatory Compliance

Management Practice	Example I&T Risk-specific Metrics
MEA03.02 Optimize response to external requirements. Review and adjust policies, principles, standards, procedures and methodologies to ensure that legal, regulatory and contractual requirements are addressed and communicated. Consider adopting and adapting industry standards, codes of good practice, and good practice guidance.	a. Average time between identifying an I&T risk issue and implementing mitigation measures b. Percent of changes in local and international laws impacting the I&T risk management posture

I&T Risk-specific Activities (in addition to COBIT 2019 core-model activities)	Capability Level
1. Incorporate new and changed regulatory obligations into the risk management activities and response plans.	3

Related Guidance (Standards, Frameworks, Compliance Requirements)	Detailed Reference
King IV Report on Corporate Governance for South Africa, 2016	Part 5.4: Governance functional areas - Principle 13

A. Component: Process *(cont.)*		
Management Practice	**Example I&T Risk-specific Metrics**	
MEA03.03 Confirm external compliance. Confirm compliance of policies, principles, standards, procedures and methodologies with legal, regulatory and contractual requirements.	a. Percent of I&T risk management practices that satisfy external compliance requirements b. Percent of noncompliance I&T risk issues identified	
I&T Risk-specific Activities (in addition to COBIT 2019 core-model activities)		**Capability Level**
1. Collect and analyze compliance data in alignment with the risk treatment plan.		3
2. Take into account the compliance level when calculating and reporting the residual risk.		
Related Guidance (Standards, Frameworks, Compliance Requirements)	**Detailed Reference**	
No related guidance for this management practice		
Management Practice	**Example I&T Risk-specific Metrics**	
MEA03.04 Obtain assurance of external compliance. Obtain and report assurance of compliance and adherence with policies, principles, standards, procedures and methodologies. Confirm that corrective actions to address compliance gaps are closed in a timely manner.	a. Time required to comply with laws and regulations	
I&T Risk-specific Activities (in addition to COBIT 2019 core-model activities)		**Capability Level**
1. Obtain and assess assertions from third-party I&T service providers on their level of compliance. Determine whether they meet enterprise risk tolerance levels.		3
2. Take into account the third-party service providers' compliance level when calculating and reporting the residual risk.		
Related Guidance (Standards, Frameworks, Compliance Requirements)	**Detailed Reference**	
CMMI Data Management Maturity Model, 2014	Supporting Processes - Process Quality Assurance	
ISO/IEC 27002:2013/Cor.2:2015(E)	18. Compliance	

C. Component: Information Flows and Items				
Management Practice	**I&T Risk-specific Inputs**		**I&T Risk-specific Outputs**	
	From	**Description**	**Description**	**To**
MEA03.01 Identify external compliance requirements.	Outside COBIT	Risk management standards and regulations	Legally required risk management plan	IRFA APO01.09 IRFA APO14.08 IRFA MEA03.02
			Report of vendor compliance breaches	IRFA APO10.05
MEA03.02 Optimize response to external requirements.	IRFA MEA03.01	External risk management compliance requirements	Updated external requirements	IRFA MEA03.03
	Outside COBIT	Applicable regulations		
MEA03.03 Confirm external compliance.	IRFA MEA03.02	Updated external requirements	Risk management compliance report	Internal
MEA03.04 Obtain assurance of external compliance.	There are no I&T risk-specific inputs for this practice.		Compliance assurance report	Internal
Related Guidance (Standards, Frameworks, Compliance Requirements)	**Detailed Reference**			
No related guidance for this component				

Domain: Monitor, Evaluate and Assess Management Objective: MEA04 — Managed Assurance	Focus Area: I&T Risk
Description	
Plan, scope and execute assurance initiatives to comply with internal requirements, laws, regulations and strategic objectives. Enable management to deliver adequate and sustainable assurance in the enterprise by performing independent assurance reviews and activities.	
Purpose	
Enable the organization to design and develop efficient and effective assurance initiatives, providing guidance on planning, scoping, executing and following up on assurance reviews, using a road map based on well-accepted assurance approaches.	
I&T Risk Focus Area Relevance	
Providing assurance is a integral part of I&T risk management.	

A. Component: Process		
Management Practice	**Example I&T Risk-specific Metrics**	
MEA04.01 Ensure that assurance providers are independent and qualified. Ensure that the entities performing assurance are independent from the function, groups or organizations in scope. The entities performing assurance should demonstrate an appropriate attitude and appearance, competence in the skills and knowledge necessary to perform assurance, and adherence to codes of ethics and professional standards.	a. Percent/frequency of independent assurance reviews of I&T risk function b. Number of information security assurance reviews c. Number of audits performed	
I&T Risk-specific Activities (in addition to COBIT 2019 core-model activities)		**Capability Level**
1. Ensure that regular assurance reviews of the I&T risk function are performed.		3
Related Guidance (Standards, Frameworks, Compliance Requirements)	**Detailed Reference**	
HITRUST CSF version 9, September 2017	06.03 Information System Audit Considerations	
Management Practice	**Example I&T Risk-specific Metrics**	
MEA04.02 Develop risk-based planning of assurance initiatives. Determine assurance objectives based on assessments of the internal and external environment and context, the risk of not achieving enterprise goals, and the opportunities associated achievement of the same goals.	a. Frequency of assurance reviews	
I&T Risk-specific Activities (in addition to COBIT 2019 core-model activities)		**Capability Level**
1. Perform risk-based audits and information security assurance activities.		2
Related Guidance (Standards, Frameworks, Compliance Requirements)	**Detailed Reference**	
King IV Report on Corporate Governance for South Africa, 2016	Part 5.4: Governance functional areas—Principle 15	
Management Practice	**Example I&T Risk-specific Metrics**	
MEA04.03 Determine the objectives of the assurance initiative. Define and agree with all stakeholders on the objectives of the assurance initiative.	a. Number of disputes on assurance objectives	
I&T Risk-specific Activities (in addition to COBIT 2019 core-model activities)		**Capability Level**
Additional I&T risk-specific activities have not been identified for this practice. The COBIT 2019 core activities are applicable.		N/A
Related Guidance (Standards, Frameworks, Compliance Requirements)	**Detailed Reference**	
CMMI Data Management Maturity Model, 2014	Supporting Processes - Process Quality Assurance	

A. Component: Process *(cont.)*		
Management Practice	**Example I&T Risk-specific Metrics**	
MEA04.04 Define the scope of the assurance initiative. Define and agree with all stakeholders on the scope of the assurance initiative, based on the assurance objectives.	a. Number of escalated scope disagreements	
I&T Risk-specific Activities (in addition to COBIT 2019 core-model activities)		**Capability Level**
1. Define practices to validate control design and outcomes and determine if the level of effectiveness supports acceptable risk (required by organizational or process risk assessment).		4
2. Where control effectiveness is not acceptable, define practices to identify residual risk (in preparation for reporting).		
Related Guidance (Standards, Frameworks, Compliance Requirements)	**Detailed Reference**	
CMMI Cybermaturity Platform, 2018	TP.LA Apply Logging and Audit Processes	
Management Practice	**Example I&T Risk-specific Metrics**	
MEA04.05 Define the work program for the assurance initiative. Define a detailed work program for the assurance initiative, structured according to the management objectives and governance components in scope.	Additional I&T risk-specific metrics have not been identified for this practice. The COBIT 2019 core metrics are applicable.	
I&T Risk-specific Activities (in addition to COBIT 2019 core-model activities)		**Capability Level**
Additional I&T risk-specific activities have not been identified for this practice. The COBIT 2019 core activities are applicable.		N/A
Related Guidance (Standards, Frameworks, Compliance Requirements)	**Detailed Reference**	
No related guidance for this management practice		
Management Practice	**Example I&T Risk-specific Metrics**	
MEA04.06 Execute the assurance initiative, focusing on design effectiveness. Execute the planned assurance initiative. Validate and confirm the design of the internal controls in place. Additionally, and specifically in internal audit assignments, consider the cost-effectiveness of the governance component design.	a. Number of design defects identified	
I&T Risk-specific Activities (in addition to COBIT 2019 core-model activities)		**Capability Level**
Additional I&T risk-specific activities have not been identified for this practice. The COBIT 2019 core activities are applicable.		N/A
Related Guidance (Standards, Frameworks, Compliance Requirements)	**Detailed Reference**	
ISF, The Standard of Good Practice for Information Security 2016	SI1 Security Audit	
ISO/IEC 27001:2013/Cor.2:2015(E)	9.2 Internal audit	

A. Component: Process *(cont.)*		
Management Practice	**Example I&T Risk-specific Metrics**	
MEA04.07 Execute the assurance initiative, focusing on operating effectiveness. Execute the planned assurance initiative. Test whether the internal controls in place are appropriate and sufficient. Test the outcome of the key management objectives in scope of the assurance initiative.	Additional I&T risk-specific metrics have not been identified for this practice. The COBIT 2019 core metrics are applicable.	
I&T Risk-specific Activities (in addition to COBIT 2019 core-model activities)		**Capability Level**
Additional I&T risk-specific activities have not been identified for this practice. The COBIT 2019 core activities are applicable.		N/A
Related Guidance (Standards, Frameworks, Compliance Requirements)	**Detailed Reference**	
ISF, The Standard of Good Practice for Information Security 2016	SI1 Security Audit	
SO/IEC 27001:2013/Cor.2:2015(E)	9.2 Internal audit	
Management Practice	**Example I&T Risk-specific Metrics**	
MEA04.08 Report and follow up on the assurance initiative. Provide positive assurance opinions, where appropriate, and recommendations for improvement relating to identified operational performance, external compliance and internal control weaknesses.	a. Number of applied recommendations	
I&T Risk-specific Activities (in addition to COBIT 2019 core-model activities)		**Capability Level**
1. Produce and issue signed-off risk management reports that reflect the agreed-upon risk response.		3
Related Guidance (Standards, Frameworks, Compliance Requirements)	**Detailed Reference**	
No related guidance for this management practice		
Management Practice	**Example I&T Risk-specific Metrics**	
MEA04.09 Follow up on recommendations and actions. Agree on, follow up and implement the identified recommendations for improvement.	a. Number of recurring I&T risk management issues b. Number of I&T risk management issues identified and resolved c. Number of unsolved high I&T risk issues	
I&T Risk-specific Activities (in addition to COBIT 2019 core-model activities)		**Capability Level**
1. Implement the necessary actions to resolve risk-related weaknesses and gaps identified.		2
Related Guidance (Standards, Frameworks, Compliance Requirements)	**Detailed Reference**	
No related guidance for this management practice		

C. Component: Information Flows and Items				
Management Practice	**I&T Risk-specific Inputs**		**I&T Risk-specific Outputs**	
	From	**Description**	**Description**	**To**
MEA04.01 Ensure that assurance providers are independent and qualified.	There are no I&T risk-specific inputs for this practice.		Competence in skills and knowledge	Internal
MEA04.02 Develop risk-based planning of assurance initiatives.	There are no I&T risk-specific inputs or outputs for this practice.			
MEA04.03 Determine the objectives of the assurance initiative.	There are no I&T risk-specific inputs or outputs for this practice.			
MEA04.04 Define the scope of the assurance initiative.	There are no I&T risk-specific inputs or outputs for this practice.			
MEA04.05 Define the work program for the assurance initiative.	There are no I&T risk-specific inputs or outputs for this practice.			
MEA04.06 Execute the assurance initiative, focusing on design effectiveness.	There are no I&T risk-specific inputs or outputs for this practice.			
MEA04.07 Execute the assurance initiative, focusing on operating effectiveness.	There are no I&T risk-specific inputs or outputs for this practice.			
MEA04.08 Report and follow up on the assurance initiative.	IRFA MEA02.04	Assessment results and remedial actions	Risk report and recommendations	Internal
MEA04.09 Follow up on recommendations and actions.	There are no I&T risk-specific inputs or outputs for this practice.			
Related Guidance (Standards, Frameworks, Compliance Requirements)	**Detailed Reference**			
No related guidance for this component				

APPENDIX A
I&T Risk Scenarios

The use of risk scenarios can enhance the I&T risk management effort by helping the risk team to understand and explain risk to stakeholders. One of the challenges for I&T risk management is to identify the important and relevant risk among all that can possibly go wrong with I&T given the pervasive presence of I&T within an enterprise and its dependence upon it.

The following table describes example generic and specific risk scenarios. It is intended to help the risk practitioner understand and predict potential situations in order to develop the appropriate holistic risk response.[13]

Figure A.1—Example Generic and Specific Risk Scenarios

	I&T Risk Scenario Category		Example Generic I&T Risk Scenarios	Example Specific I&T Risk Scenarios
1	**IT investment decision making, portfolio definition and maintenance**	A	Programs selected for implementation are misaligned with corporate strategy and priorities.	• Programs selected for implementation are misaligned with corporate strategy and priorities, resulting in an inappropriate new customer relationship management (CRM) application that fails to support the customer service strategy.
		B	IT-related investments are not supporting the digital strategy of the enterprise.	• The enterprise has a misaligned investment in a new accounting system that limits the investment funds available for the multichannel customer-facing application.
		C	Wrong software (cost, performance, features, compatibility, redundancy, etc.) is selected for acquisition and implementation.	• Incompatible software for older production machines is implemented to replace legacy systems, resulting in loss of production and business.
		D	Wrong infrastructure (cost, performance, features, compatibility, etc.) is selected for implementation.	• An investment in on-premise infrastructure is made despite lacking requisite capabilities to successfully implement and support, resulting in increased costs and service failures. • An investment in cloud infrastructure is made that results in limited local availability necessary to support business goals, resulting in service availability failures.
		E	Duplication or important overlaps among different investment initiatives exist.	• Two large departments in the enterprise are heavily investing in new customer-facing applications but both are developing their own (different) customer database, resulting in increased organizational costs.
		F	New investment programs create long-term incompatibility with the enterprise architecture.	• Programs selected for implementation are misaligned with the enterprise architecture, resulting in an inappropriate new accounting application that fails to integrate with the current application landscape, resulting in increased costs and service failures.
		G	Competing resources are allocated and managed inefficiently and are misaligned to business priorities.	• Technical experts in critical technologies are assigned to work on less critical projects instead of high-priority and key projects, resulting in product and service delivery failures.

[13] These scenarios were first published by ISACA in *Risk IT Practitioner Guide, 2nd Edition*, USA, 2020, *www.isaca.org/bookstore/bookstore-risk-print/ritpg2.*

Figure A.1—Example Generic and Specific Risk Scenarios *(cont.)*

I&T Risk Scenario Category		Example Generic I&T Risk Scenarios		Example Specific I&T Risk Scenarios
1	**IT investment decision making, portfolio definition and maintenance** *(cont.)*	H	Shadow IT and Software as a Service (SAAS) solutions are deployed from departmental budgets, generating longer-term architectural problems and excessive overall IT cost.	• The marketing department acquires its own SAAS-based customer management system, resulting in expensive additional developments of adequate interfaces with legacy systems.
		I	Inadequate consideration is given to future requirements or demands.	• The criteria for IT investment selection do not include future user requirements, leading to incomplete solutions or solutions not aligned with business strategy.
		J	Initiative economics are not considered during investment decisions.	• Project economics are not adequately used for investment selection, leading to high-return projects being discarded or negative return projects not abandoned.
2	**Program and projects life cycle management**	A	Failing projects (cost explosion, excessive delays, scope creep, changed business priorities) are not terminated by senior management.	• A long-running and ambitious project to build a new human resources (HR) system has not yet delivered any tangible results in its attempt to integrate payroll management, time management and performance management. The enterprise, despite the lack of results, keeps (under)funding the project for the next period.
		B	IT projects exceed planned budgets.	• A major project for a new enterprise resource planning (ERP) implementation approaches a 100 percent budget overrun.
		C	IT projects do not meet business requirements.	• The new CRM system is performing unacceptably slowly and is unstable because of many crashes, resulting in an availability rate of less than 90 percent. • The new CRM system's implementation of privacy regulations fails for 50 percent of the test cases during acceptance testing, creating a noncompliance exposure. • The new CRM system's functionality and user interface (UI) are not accepted by the intended users, due to not obtaining the active involvement throughout the program/project life cycle of all stakeholders (including the sponsor). • Because of incomplete business requirements analysis and/or functional design, the new HR system is not in line with users' expectations. • Because of an ineffective testing process, the new ERP system contains many bugs and impacts productivity.
		D	IT projects are delivered late.	• The enterprise's new website, including the online shop, is delivered six months behind schedule, creating a missed market opportunity.
		E	New application software is not adopted by users.	• Incorrect requirements analysis results in users' failure to adopt the new CRM application and a loss of efficiency.
		F	Immature software (early adopters, bugs, etc.) is implemented.	• Business application experiences service disruption due to software defects.
		G	Security requirements are inadequately assessed.	• The lack of consideration for security requirements leads to an insecure solution for a new remote access application, requiring extensive rework and delays.
		H	Program/project scope is not well managed.	• Frequent and late program scope changes forced by the system owner upon the developer lead to unstable and inconsistent solutions, causing extensive rework and delays.

Figure A.1—Example Generic and Specific Risk Scenarios *(cont.)*

I&T Risk Scenario Category		Example Generic I&T Risk Scenarios		Example Specific I&T Risk Scenarios
2	**Program and projects life cycle management** *(cont.)*	I	Inadequate KPIs exist to measure program purpose achievement.	• The enterprise is unable to manage projects correctly because it is unaware of its true state, leading to late/wrong project decisions, delays and extensive overruns.
3	**IT cost and oversight**	A	There is extensive dependency on and use of user-created and ad hoc solutions.	• Extensive dependency on the use of end-user computing for important information needs leads to security deficiencies. • Extensive dependency on the use of end-user computing for important information needs leads to inaccurate and unreliable data. • Extensive dependency on ad hoc solutions for important information needs leads to inefficient use of resources and additional cost.
		B	Change management over ad hoc solutions is inadequate.	• Inadequate change management and quality control over ad hoc user solutions leads to wrong computing results and business decisions (e.g., business case outcomes are wildly inaccurate).
		C	Cost and ineffectiveness are related to I&T-related purchases outside of the I&T procurement process.	• Different departments are acquiring their own office automation solutions. • Different departments are entering into agreements with different outsourcing vendors or service providers.
		D	Inadequate requirements lead to ineffective service level agreements (SLAs).	• Relevant stakeholders did not participate in the requirements phase, resulting in incomplete requirements and incomplete SLAs.
		E	There is a lack of funding for I&T-related investments.	• Insufficient funding for necessary security updates leads to failure of IT systems through cyberattacks. • Lack of funding for new I&T innovations results in a slower go-to-market response and a loss of competitive advantage.
4	**IT expertise, skills and behavior**	A	There is a lack or mismatch of IT-related skills within IT (e.g., due to new technologies or working methods).	• Current IT staff do not have the required skills for new technologies (e.g., blockchain), resulting in longer learning curves and delays in projects. • Current IT operations and development staff are not trained in DevOps tools and ways of working, resulting in less-than-anticipated benefits of the DevOps approach.
		B	A lack of business understanding by IT staff affects service delivery/projects quality.	• A lack of business understanding by IT staff affects the quality of the new ERP system due to deficient implementation of user requirements and implementation of unnecessary features.
		C	The enterprise is unable to recruit and retain IT staff.	• The HR department chronically fails to recruit sufficiently skilled information security experts, leading to increased exposure and increased costs for external experts. • There is an insufficient return on investment regarding training due to early departure of trained IT staff (e.g., individuals with advanced business degrees).
		D	Individuals with unsuitable profiles are recruited because of lack of due diligence in the recruitment process.	• The HR department recruits employees with false credentials, which results in damage to the brand in the long run. • A failure to perform the necessary due diligence on a new hire resulted in insider hacking of confidential data.

Figure A.1—Example Generic and Specific Risk Scenarios *(cont.)*

I&T Risk Scenario Category		Example Generic I&T Risk Scenarios		Example Specific I&T Risk Scenarios
4	**IT expertise, skills and behavior** *(cont.)*	E	I&T training is lacking.	• The lack of I&T training leads to the departure of IT staff. • The lack of I&T-related training leads to increased quality problems of delivered services. • The enterprise is unable to update the I&T skills to the proper level through training.
		F	There is an overreliance on key staff for I&T services.	• Critical in-house knowledge is lost when key personnel leave the organization.
		G	No formal staff performance review exists.	• The lack of a formal performance review of IT development staff leads to subpar software quality and delayed delivery time of new applications.
		H	Duties are ineffectively segregated.	• Ineffective segregation of duties for the purchasing application allows for fraudulent transactions.
		I	Communication challenges exist between IT and users.	• Difficult relationships between IT and users lead to poorly understood user requests and deficient solutions.
5	**Enterprise/IT architecture**	A	Enterprise architecture is complex and inflexible, obstructing further evolution and expansion, leading to missed business opportunities.	• A new customer relationship management application cannot be implemented or has to be reduced in functionality because communication with other internal systems is too slow because of very slow interfaces.
		B	The enterprise fails to adopt and exploit new infrastructure or abandon obsolete infrastructure on a timely basis.	• Abandoned infrastructure cabling reduces the efficiency of heating, venting and air conditioning (HVAC) systems.
		C	The enterprise fails to adopt existing enterprise/IT architecture when designing and implementing new technology solutions.	• The marketing department decides to use a SAAS solution without consulting the enterprise architects, resulting in a solution that is incompatible with current internal systems, requiring the expensive build of additional interfacing software.
		D	The enterprise fails to adopt and exploit new software (functionality, optimization, etc.) or abandon obsolete applications on a timely basis.	• The crash of an older, unsupported accounting application results in high profit loss. • The failure to abandon obsolete applications takes up unnecessary time and resources (i.e., money and people).
		E	Undocumented enterprise architecture leads to inefficiencies and duplications.	• The marketing department decides to use a SAAS solution without consulting the enterprise architects, resulting in a solution that is incompatible with current internal systems, requiring the expensive build of additional interfacing software.
		F	There is an excessive number of exceptions on enterprise architecture standards.	• Over the previous year, almost 75 percent of the new projects have requested (and been granted) exceptions to the enterprise architecture standards. This renders the standards in practice irrelevant and will create inconsistent solutions and excessive cost going forward.
6	**IT operations**	A	Errors are made by IT staff (during backup, upgrades of systems, maintenance of systems, etc.)	• IT operations staff (internal or at the service provider) enters a wrong command during an upgrade of a system, leaving the system vulnerable, from a security point of view.

Figure A.1—Example Generic and Specific Risk Scenarios *(cont.)*

I&T Risk Scenario Category		Example Generic I&T Risk Scenarios		Example Specific I&T Risk Scenarios
6	**IT operations** *(cont.)*	B	Incorrect information is input by IT staff or system users.	• The system manager of a critical server enters incorrect network information into the system, crippling the communication speed of the system to unacceptable levels.
		C	Backup/recovery management is deficient and includes mislabeling or misplacing backup media.	• Deficient backup/recovery management is causing data loss when backups cannot be restored after a hardware incident. • Deficient backup/recovery management is causing data loss when backups cannot be restored after a ransomware incident. • Backup status is not properly monitored and/or backup issues are not being resolved in a timely manner, leading to potential data loss when a restore would be needed.
		D	Patch/vulnerability management is inadequate.	• Inadequate patch and vulnerability management leaves systems vulnerable to attacks or crashes, reducing service levels.
		E	Performance and operations monitoring is inadequate.	• Inadequate systems or application monitoring can cause system degradation to remain unnoticed for too long, leading to service interruptions.
		F	Facilities' resilience is inadequate.	• Power interruptions or surges without an adequate Unterruptible Power Supply system damage computing equipment and can cause service interruptions or data loss.
7	**User access rights management**	A	Software is tampered with.	• An IT operations manager in a DevOps environment makes a number of software changes he believes will make a new application run smoother inadverntly causing an error or a system performance issue.
		B	Software is intentionally modified or manipulated, leading to wrong data.	• The modification of database source code by a developer with wrong intentions results in wrong data.
		C	The security configuration is intentionally modified or manipulated.	• Tampering with the security configuration of network equipment creates important vulnerabilities, allowing attacks to be successful.
		D	Software is intentionally modified or manipulated, leading to fraudulent actions.	• A software engineer makes unauthorized changes to a payment system so he/she can make unauthorized and fraudulent payments.
		E	Software is unintentionally modified, leading to inaccurate results.	• Unintentional false modification of database source code by an unaware developer results in loss of production and data.
		F	Unintentional errors are made in configuration and change management.	• Change management or configuration errors may cause system outages, leading to a service interruption of customer-facing applications.
		G	Unintentional communications are issued.	• An email with sensitive personal information is sent to the wrong recipients, leading to breach of privacy claims.
		H	Access rights from prior roles are abused to access IT infrastructure.	• A lack of internal control of assigned roles leads to improper access rights of users. Financial data are accessible and stolen by employees.
		I	Authorized users intentionally or accidentally conduct unauthorized actions.	• A privileged user is accessing private personal information from clients for no valid business reason, exposing the enterprise to a privacy violation.
		J	Access rights or passwords are shared with others.	• Authorized users (privileged or not) share their credentials with other users, allowing these other users actions to which they are not entitled and obstructing nonrepudiation.

Figure A.1—Example Generic and Specific Risk Scenarios *(cont.)*

	I&T Risk Scenario Category		Example Generic I&T Risk Scenarios	Example Specific I&T Risk Scenarios
7	**User access rights management** *(cont.)*	K	Privileged/emergency account management is inadequate.	• Deficient management of privileged accounts results in an excessive number of users with privileges, increasing the likelihood of unauthorized actions and reducing nonrepudiation.
		L	User access provisioning is ineffective.	• Policy dispensation processes during a disaster are ineffective. (During a disaster, some existing controls may not be feasible. Thus, there should be a dispensation process to assess the risk and track for the dispensation.)
		M	The access matrix is not defined or implemented.	• A poorly designed or implemented access matrix may result in users with (typically) excessive privileges, resulting in poorly implemented business controls over (e.g., financial) transactions.
		N	There is a conflict of interest during user access management.	• Pressure from IT management on the security team under its authority results in excessive access rights been given to some users.
8	**Software adoption and use**	A	Users fail to adopt new application software.	• Users refuse or are reluctant to use a new software application because of lack of training/communication, thereby foregoing efficiency gains.
		B	Users use new software inefficiently.	• Users are provided with a new system for managing customer complaints that automates the workflow, but they keep emailing directly to the support persons in parallel.
		C	Unintended use of new software applications appears.	• Business users are not using a new software application as intended, leading to productivity losses.
		D	Enterprise users are unable to use the software to realize desired outcomes (e.g., required business model or organizational changes are not made).	• A new business application experiences an outage due to a lack of familiarity with a newly deployed server system.
		E	Operational glitches arise when new software is made operational.	• The organization's public-facing website is intermittently inaccessible following the deployment of a software upgrade.
		F	Critical application software malfunctions regularly.	• A critical customer-facing application becomes very unstable after the latest software change, sharply reducing the application's availability and causing numerous customer complaints.
		G	Application software is obsolete (old technology, poorly documented, expensive to maintain, difficult to extend, not integrated in current architecture, etc.).	• A key business application is unstable due to older, unsupported technologies, resulting in periodic outages.
		H	The enterprise is unable to revert to former versions of software when operational issues arise with the new version.	• A key business application is disabled due to a failure in the new version of the software, and the enterprise is unable to revert to the older version.
		I	A software-induced corrupted data(base) leads to inaccessible data.	• A bug in the latest release of software results in corruption of data in a key business application, which makes the application unavailable.
		J	Software-induced information integrity issues appear.	• A software change introduces new business and processing logic, but errors are causing information integrity problems, leading to wrong business decisions (e.g., credit approvals, customer order processing).

Figure A.1—Example Generic and Specific Risk Scenarios *(cont.)*

I&T Risk Scenario Category		Example Generic I&T Risk Scenarios		Example Specific I&T Risk Scenarios
9	**IT hardware**	A	New infrastructure is installed and, as a result, systems become unstable, leading to operational incidents (e.g., a bring your own device [BYOD] program).	• Outage of a key business application results from the instability of new Cisco network hardware.
		B	Systems cannot handle transaction volumes when user volumes increase.	• A database is unable to handle transaction volumes, resulting in periodic outages. • An application is unable to handle transaction volumes, resulting in slow response times for customers.
		C	Systems cannot handle system load when new applications or initiatives are deployed.	• Network throughput constraints introduce transaction delays when the new firewall is deployed.
		D	Utilities (e.g., telecom, electricity) fail.	• Online consumer transactions are interrupted when a storm causes an electricity service disruption.
		E	Hardware fails due to overheating and/or other environmental conditions, such as humidity.	• Online retail transactions are interrupted when the system hardware fails due to HVAC failure.
		F	Hardware fails due to a lack of preventive maintenance.	• An HVAC system fails because of late maintenance, causing a computer room outage.
		G	Hardware components are damaged, leading to (partial) destruction of data by internal staff.	• A disgruntled staff member physically destroys a hard drive, leading to data loss.
		H	Portable media containing sensitive data (CDs, USB drives, portable disks, etc.) are lost or disclosed.	• Customer information is compromised due to employees using unauthorized and unencrypted USB drives.
		I	The enterprise experiences extended resolution time or support delays in cases of hardware incidents.	• Customers experience lengthy service delays due to outages of hardware that is no longer supported by the vendor.
		J	Hardware components are configured erroneously.	• After an upgrade of the storage system, several disks are incorrectly configured, making the applications using the data run erroneously.
		K	Hardware commissioning and decommissioning management is inadequate.	• A deficient hardware decommissioning procedure causes data to not be destroyed before the equipment is decommissioned and removed from the premises.
		L	Hardware configuration compliance management is inadequate.	• An operator's failure to comply with a hardware configuration procedure leaves a new piece of equipment poorly configured, causing it to be exploited and generating a service interruption.
		M	Vendor support comes to an end.	• The vendor of a legacy system used by the enterprise has ceased all support for the system, resulting in total service interruption in case of a hardware defect.
10	**Internal and external security threats (hacker, malware, etc.)**	A	Unauthorized (internal) users successfully break into systems.	• An internal user breaks into a key business application and deletes customer account information.
		B	Service interruption occurs due to a denial of service (DoS) attack.	• A distributed denial of service (DDoS) attack results in performance degradation of a key business website.
		C	A website is defaced.	• A flaw in the security of the main marketing website results in a website defacement that creates embarrassment for the enterprise.

Figure A.1—Example Generic and Specific Risk Scenarios *(cont.)*

I&T Risk Scenario Category		Example Generic I&T Risk Scenarios		Example Specific I&T Risk Scenarios
10	**Internal and external security threats (hacker, malware, etc.)** *(cont.)*	D	The enterprise experiences a malware attack.	• Critical operational servers are intruded by malware. • Laptops are regularly infected with malware. • The enterprise is forced to pay ransom due to a set of DDoS attacks.
		E	Industrial espionage takes place.	• A foreign state or enterprise hacks the enterprise's systems and obtains important and confidential product and client information, reducing the enterprise's competitive position.
		F	Hacktivism occurs.	• A DDoS attack is perpetrated by opponents of the company's environmental policies, taking the website offline for more than 24 hours.
		G	A disgruntled employee implements a time bomb, which leads to data loss.	• Data in the central customer information database are deleted and made unrecoverable by a disgruntled employee who implements a data deletion time bomb.
		H	Company data are stolen through unauthorized access gained by a phishing attack.	• Employees use unauthorized USB devices, distributed on the parking lot, resulting in a loss of data. • An unaware employee opens a phishing email, resulting in unauthorized access to financial data.
		I	A foreign government issues attacks on critical systems.	• A foreign government attacks critical systems. • Organized crime groups intrude into systems for fraudulent transactions.
		J	The data center is destroyed by staff (sabotage, etc.).	• Operations staff of the data center plant a bomb in the HVAC facilities.
		K	A device with sensitive data is stolen.	• An authorized administrator steals a device with sensitive financial data.
		L	IT equipment is accidentally damaged.	• The cleaning staff accidentally damage network cabinet equipment. • IT operations staff accidentally damage disk racks.
		M	A key infrastructure component is stolen.	• A staff member of the data center cleaning crew steals the laptop managing the access control system to the data center, making it impossible to securely enter the data center.
		N	Hardware (security devices, etc.) is intentionally tampered with.	• A technician tampers with the fingerprint recognition system for sensitive premises so the system allows anybody in.
11	**Third-party/ supplier incidents**	A	An outsourcer performs inadequately in a large-scale, long-term outsourcing arrangement (e.g., through lack of supplier due diligence regarding financial viability, delivery capability and sustainability of supplier's service).	• Delivery of key product components is interrupted due to the financial failure of an outsourced provider.
		B	The enterprise accepts unreasonable terms of business of IT suppliers (e.g., through lack of legal advice).	• Cost overruns occur due to poor negotiation of terms in agreements with IT service providers.
		C	Vendors deliver inadequate support and services, not in line with the SLA.	• Delays in software fixes to the online retail website occur due to software vendors who fail to meet SLA requirements for timeliness and/or quality.

Figure A.1—Example Generic and Specific Risk Scenarios *(cont.)*					
I&T Risk Scenario Category		Example Generic I&T Risk Scenarios		Example Specific I&T Risk Scenarios	
11	**Third-party/ supplier incidents** *(cont.)*	D	Noncompliance with software license agreements (use and/or distribution of unlicensed software, etc.) occurs.	•	Legal action is brought by a software provider who identifies that the enterprise has violated the software license agreement by unmanaged distribution of the software.
		E	The enterprise is unable to transfer to alternative suppliers due to overreliance or overdependence on the current supplier.	•	Product delivery services are degraded due to overreliance on a single service provider that cannot scale up to meet increased demand.
		F	IT services (especially cloud services) are purchased by the business without consultation/involvement of IT, resulting in an inability to integrate the service with in-house services.	•	A cloud service provider is selected that is incompatible with internal applications.
		G	Penalties are incurred due to noncompliance with SLAs.	•	Legal action is brought against the enterprise for failure to meet contractual requirements for delivery of services.
		H	The SLA is inadequate to obtain required services.	•	An inadequate SLA does not describe required service levels, which consequently are not delivered by the third-party provider, affecting staff productivity.
		I	Service is terminated.	•	A service provider ceases to exist due to bankruptcy, leading to a service interruption.
		J	Operational service is interrupted by cloud providers.	•	The cloud provider that hosts the enterprise's customer database has an operational malfunction of its systems, resulting in unavailability of the customer database for a day.
		K	Noncompliance with security requirements occurs.	•	A cloud service provider is not compliant with contractual security requirements, which leads to a confidentiality breach with legal consequences for the enterprise.
		L	Third-party outsourcers do not deliver projects as per contractual agreements (any combination of exceeded budgets, quality problems, missing functionality, late delivery).	•	The contractor for the new ERP system fails to deliver the new system by the promised deadline and will probably deliver one year late.
		M	Monitoring of the SLA with third-party providers is inadequate.	•	Due to a lack of monitoring of the performance of a cloud provider, degradation of services and noncompliance with contractual service levels remain unnoticed, affecting staff productivity.
		N	The cloud provider loses data.	•	The cloud provider that hosts the enterprise's customer database has a major malfunction of its systems, resulting in a loss of the previous month's customer updates.

Figure A.1—Example Generic and Specific Risk Scenarios *(cont.)*

I&T Risk Scenario Category			Example Generic I&T Risk Scenarios	Example Specific I&T Risk Scenarios
12	Noncompliance	A	Noncompliance with local or international regulations (e.g., privacy, accounting, manufacturing, environmental) occurs.	• The enterprise experiences high fines and major damage to its brand due to unnecessary mistakes made against GDPR regulations. • Inadequate analysis of regulations leads to unintended noncompliance and, as a result, fines are incurred.
		B	The enterprise is unaware of potential regulatory changes that may have a business impact.	• The enterprise undertakes incomplete follow-up of changes in environmental regulations.
		C	The enterprise experiences operational obstacles caused by regulations.	• The regulator prevents cross-border data flow due to insufficient controls.
		D	The enterprise experiences compliance failures with internal procedures.	• Legal action is brought against the organization by personnel who cite a failure to comply with internal safety policies. • Users do not comply with internal policies on downloading external software, leading to infected systems and subsequent service interruptions.
		E	Cross-border regulations cause obstacles for the enterprise.	• The existence of different regulations between countries prohibits or inhibits information transfer, leading to additional costs and efficiency loss.
13	Geopolitical issues	A	The enterprise experiences service interruptions due to disruptive incidents (e.g., physical attack) in foreign premises.	• The data center in a foreign country is bombed and causes important supply chain problems.
		B	Government interference and national policies impact the enterprise.	• Government interference and national policies limit international service or product delivery.
		C	The enterprise experiences targeted action from government-sponsored groups or agencies.	• Targeted action against the enterprise results in destruction of critical infrastructure. • Targeted action from foreign entities obtains confidential product information, damaging the competitive position of the enterprise.
14	Industrial action	A	Facilities and buildings are not accessible because of a labor union strike.	• A reduction in productivity occurs because laborers at the production plant are unable to go to work due to a union strike.
		B	Third-party providers are not able to provide services because of a strike.	• External technicians cannot provide technical services because the premises are blocked by strikers, leading to service degradation.
		C	Key staff are not available as a result of external industrial action (e.g., a transportation or utilities strike).	• A reduction in productivity occurs because key staff are unable to go to work due to a transportation strike.

Figure A.1—Example Generic and Specific Risk Scenarios *(cont.)*

	I&T Risk Scenario Category		Example Generic I&T Risk Scenarios	Example Specific I&T Risk Scenarios
15	**Acts of nature**	A	An earthquake destroys or damages important IT infrastructure.	• An earthquake damages the IT operations center, resulting in the outage of online retail web services.
		B	A tsunami destroys critical premises.	• A tsunami destroys a major distribution center, resulting in an inability to meet product delivery deadlines.
		C	Major storms and tropical cyclones damage critical infrastructure.	• A major storm damages electrical infrastructure, resulting in the unavailability of grid-based power to the data center for more than two weeks.
		D	Major wildfires and/or flooding damage critical premises.	• A wildfire forces staff of a data center to abandon the premises, causing prolonged service interruptions. • Flooding of the premises causes irreparable damage to computer equipment.
		E	Changing environmental conditions affect the enterprise.	• A rising water table makes a critical location unusable, requiring relocation. • A rising temperature makes critical locations uneconomical to operate.
		F	A pandemic affects people and the economy.	• Staff are unable to access the office premises due to natural disaster (e.g., flood or pandemic), causing efficiency degradation. • Permanent or prolonged loss of key employees during the pandemic causes service interruptions, knowledge gaps and efficiency degradation. • The enterprise is unable to acquire key resources (e.g., bandwidth) during the pandemic.
16	**Emerging technologies and innovation**	A	New and important technology trends are not identified.	• A failure to leverage advancements in technology results in a loss of market share.
		B	The enterprise fails to appreciate the value and potential (new functionality, process optimization) of new technologies.	• The enterprise incurs important opportunity costs by not adopting in a timely manner a new software platform for customer interactions.
		C	The enterprise experiences problematic early adoption of new technology.	• Early adoption of new technology causes inconsistent stability and reliability, causing unanticipated interruptions to operations, demand for massive support and frequent updates, all resulting in efficiency degradation.
		D	The enterprise fails to understand or address the risk linked to the adoption of new technology (e.g., lack of hardening guidance).	• Adoption of insufficiently hardened new technology may expose the enterprise to security risk and create service interruptions.
		E	Emerging technology disappears from the market.	• New technology that has been adopted early disappears from the market, leading to service interruptions or additional cost for replacement technologies.

Figure A.1—Example Generic and Specific Risk Scenarios *(cont.)*				
I&T Risk Scenario Category		**Example Generic I&T Risk Scenarios**		**Example Specific I&T Risk Scenarios**
17	**Environmental**	A	Equipment used by the enterprise is not environmentally friendly (e.g., power consumption, packaging).	• The use of herbicide sprayers out of accordance with established safety practices results in damage to endangered wildlife species.
		B	Equipment used by the enterprise does not comply with environmental regulations or energy consumption standards.	• Computing equipment does not comply with environmental regulations, leading to penalties.
		C	Equipment used by the enterprise does not comply with the latest energy consumption standards.	• The energy consumption by the enterprise's computing equipment is excessive and leads to higher cost.
18	**Data and information management**	A	Sensitive information is discovered by unauthorized persons due to inefficient retaining/archiving/ disposing of information.	• Accidental disclosure of sensitive client information due to failure to follow information handling guidelines (e.g., disposal of documents) leads to legal procedures.
		B	Data are subject to intentional illicit or malicious modification.	• Intentional illicit or malicious modification of accounting data leads to an unfavorable audit opinion. • Intentional illicit or malicious modification of a patient's medical records leads to legal procedures.
		C	Unauthorized disclosure of sensitive information through email or social media occurs.	• A design engineer shares new and confidential product information on his/her social media accounts with lots of followers. • An HR staff member shares confidential news on layoffs on his/her social media accounts, causing lots of press attention and major disruption in the organization.
		D	The enterprise loses intellectual property (IP) and/or suffers leakage of competitive information.	• The enterprise loses IP and/or its competitive information is leaked due to key team members leaving the organization and taking information with them. • Hacking results in a loss of IP and/or leakage of competitive information.
		E	The enterprise's information is of poor quality.	• Management of a service organization is unable to optimize its staff planning because of the poor quality of staff utilization and consistent communication.
		F	Master data management is inadequate.	• Due to poor master data management or poor data architecture, multiple instances of the same data exist on the enterprise's systems, leading to important customer service and efficiency degradation.
		G	Information classification is inadequate.	• Poor information classification leads to inadequate protection of information and disclosure of confidential marketing information, reducing the competitive advantage of the enterprise.

APPENDIX B

Risk Scenario Categories Mapped to COBIT 2019 Governance and Management Objectives

The following table shows the 18 risk categories mapped to the 40 COBIT 2019 governance and management objectives. Those categories listed with "H" have a high correlation, those listed with "M" have a medium correlation, those listed with "L" have a low correlation and those with no value have little to no correlation.

COBIT Objective	RISK CAT01 IT Investment Decision Making, Portfolio Definition & Maintenance	RISK CAT02 Program & Projects Life Cycle Management	RISK CAT03 IT Cost & Oversight	RISK CAT04 IT Expertise, Skills & Behavior	RISK CAT05 Enterprise/ IT Architecture	RISK CAT06 IT Operations	RISK CAT07 User Access Rights Management	RISK CAT08 Software Adoption & Use	RISK CAT09 Hardware Incidents	RISK CAT10 Internal & External Security Threats	RISK CAT11 Third-Party/Supplier Incidents	RISK CAT12 Noncompliance	RISK CAT13 Geopolitical Issues	RISK CAT14 Industrial Action	RISK CAT15 Acts of Nature	RISK CAT16 Technology-Based Innovation	RISK CAT17 Environmental	RISK CAT18 Data & Information Management
EDM01	H	M	H				M					H	M			M	M	M
EDM02	H	M			M							L				H	L	H
EDM03	M	M						L		M		H	H				M	H
EDM04	H		H	H	M						M	L		M			M	H
EDM05	H	L	H				M	L			L	H	H				M	M
APO01	M	H	M		M	M	H	M		H	H	H				H	M	H
APO02	M				H			M	L	L	M					M	M	L
APO03	M				H			M		M	M					M		H
APO04					L											H		
APO05	H	M	M		M			M	M							M		
APO06	M	H	H								M		M			M	M	
APO07				H		M	H	H		M			M	H		M	M	
APO08				M	M			H		M	M					H		M
APO09			M				M	H		M	H							
APO10		M	H				M	M	H	M	H	M	M					
APO11		H						H										M
APO12							H	M		H					M			
APO13							H			H		H						
APO14							H	M		M		H		M	H	M		H
BAI01		H			M			H										
BAI02	M	M			M			H		M								
BAI03		H			M			H		H								
BAI04		L																
BAI05		M		M				H										
BAI06						H	H	M		H								H
BAI07						M	H	H		M								
BAI08				M		H		H						M				M
BAI09						L	H											
BAI10						M	H	M		H								
BAI11		H																
DSS01						H	H		H	M							M	
DSS02						H	M	H	M	H								
DSS03						H	L	H		L								
DSS04						H	H		H	H		M		H	H			M
DSS05						H	H		M	H		H		H	M			H
DSS06						H	H	M		M		M						H
MEA01	L	M	M			M	M	M		H	M	M	M		M			M
MEA02	L	M	M			H	H	M		H	M	M	H		M			M
MEA03		L				L	M			H	M	H	M					M
MEA04	L	M					H	M		H	M	M	H		M	M		M

Page intentionally left blank

APPENDIX C

Template for Risk Register Entry

The following template is a sample risk register, adapted from *COBIT® 5 for Risk*.[14]

Figure C.1—Risk Register

Part I—Summary Data						
Risk statement						
Risk owner						
Date of last risk assessment						
Due date for update of risk assessment						
Risk category	☐ STRATEGIC (IT Benefit/Value Enablement)		☐ PROJECT DELIVERY (IT Program and Project Delivery)		☐ OPERATIONAL (IT Operations and Service Delivery)	
Risk classification (copied from risk analysis results)	☐ LOW	☐ MEDIUM		☐ HIGH	☐ VERY HIGH	
Risk response	☐ ACCEPT	☐ TRANSFER		☐ MITIGATE	☐ AVOID	
Part II—Risk Description						
Title						
High-level scenario (from list of sample high-level scenarios)						
Detailed scenario description—Scenario components	Actor					
	Threat Type					
	Event					
	Asset/Resource					
	Timing					
Other scenario information						
Part III—Risk Analysis Results						
Frequency of scenario (number of times per year)	0	1	2	3	4	5
	N ≤ 0,01 ☐	0,01 < N ≤ 0,1 ☐	0,1 < N ≤ 1 ☐	1 < N ≤ 10 ☐	10 < N ≤ 100 ☐	100 < N ☐
Comments on frequency						
Impact of scenario on business	0	1	2	3	4	5
1. Productivity	Revenue Loss Over One Year					
Impact rating	I ≤ 0,1% ☐	0,1% < I ≤ 1% ☐	1% < I ≤ 3% ☐	3% < I ≤ 5% ☐	5% < I ≤ 10% ☐	10% < I ☐
Detailed description of impact						
2. Cost of response	Expenses Associated With Managing the Loss Event					
Impact rating	I ≤ 10k$ ☐	10K$ < I ≤ 100K$ ☐	100K$ < I ≤ 1M$ ☐	1M$% < I ≤ 10M$ ☐	10M$ < I ≤ 100M$ ☐	100M$ < I ☐
Detailed description of impact						
3. Competitive advantage	Drop-in Customer Satisfaction Ratings					
Impact rating	I ≤ 0,5 ☐	0,5 < I ≤ 1 ☐	1 < I ≤ 1,5 ☐	1,5 < I ≤ 2 ☐	2 < I ≤ 2,5 ☐	2,5 < I ☐
Detailed description of impact						
4. Legal	Regulatory Compliance—Fines					
Impact rating	None ☐	< 1M$ ☐	< 10M$ ☐	< 100M$ ☐	< 1B$ ☐	> 1B$ ☐
Detailed description of impact						
Overall impact rating (average of four impact ratings)						
Overall rating of risk (obtained by combining frequency and impact ratings on risk map)	☐ LOW	☐ MEDIUM		☐ HIGH	☐ VERY HIGH	

[14] ISACA, *COBIT® 5 for Risk*, USA, 2013, figure 62, *www.isaca.org/bookstore/cobit-5/wcb5rk*

Figure C.1—Risk Register (*cont.*)

Part IV—Risk Response				
Risk response for this risk	☐ ACCEPT	☐ TRANSFER	☐ MITIGATE	☐ AVOID
Justification				
Detailed description of response (NOT in case of ACCEPT)	Response Action		Completed	Action Plan
	1.		☐	☐
	2.		☐	☐
	3.		☐	☐
	4.		☐	☐
	5.		☐	☐
	6.		☐	☐
Overall status of risk action plan				
Major issues with risk action plan				
Overall status of completed responses				
Major issues with completed responses				
Part V—Risk Indicators				
Key risk indicators for this risk	1. 2. 3. 4.			

Source: ISACA, *COBIT® 5 for Risk*, USA, 2013, figure 62, *www.isaca.org/bookstore/cobit-5/wcb5rk*

APPENDIX D
IT Risk Reporting Examples

IT Risk Dashboard

The following is an example of a summarized IT Risk Indicator Dashboard (**figure D.1**). The Dashboard displays three levels of indicators: business impact indicators, risk indicators and control indicators. Each level of risk indicators relate to the different level of goals and practices. The business impact indicators are related to the enterprise goals, the risk indicators are related to the alignment goals, and the control indicators are related to the management practices. The I&T risk owners and risk stakeholders may decide the period of time over which the indicators are measured, i.e., in the current month, over the last reporting period, or year to date (YTD). Risk practitioners need to determine indicator metrics that are important to the organization and its stakeholders.

Figure D.1—Example of IT Risk Indicator Dashboard

Indicator Metric	Value	Trend
Business Impact Indicators		
Monetary value of losses due to IT risk (YTD)	$30,000	→
# of customers affected by IT risk incidents (YTD)	$20,000	↗
# of regulator notifications (YTD) (figure D.2)	3	→
Risk Indicators		
# of non-compliance incidents (YTD) (figure D.3)	10	→
% of critical system availability (Current Month)	95%	↘
# of significant IT risk incidents (YTD)	50	↗
Control Indicators		
% of control environment effectiveness (Current Month) (figure D.4)	80%	↗
# of control issues (YTD) (figure D.5)	50	↗
# of overdue management actions (Current Month)	3	↘
# of policy reviews (YTD)	10	→
% of staff trained (YTD)	75%	↗

IT Risk Graphs

Risk stakeholders may require a historical view of risk indicators to identify trends and projections. The following example shows IT trending graphs that can give organizations a starting point to communicate and evaluate indicators which may be indicative of major shifts impacting the business, risk, and controls.

Figure D.2 is an example of an IT trending graph to help enterprises effectively assess the overall financial impact of regulatory fines and penalties over a 12-month time frame. Upward trends can be an early indicator of increased regulatory enforcement or perhaps highlight noncompliance areas which require attention to ensure risk is constrained to acceptable levels for the organization. The enterprise can assess whether trends are within an acceptable risk range (green), the early warning range (yellow/amber) and the unacceptable risk threshold (red).

Figure D.2—Accumulated Regulatory Fines & Penalties Metric

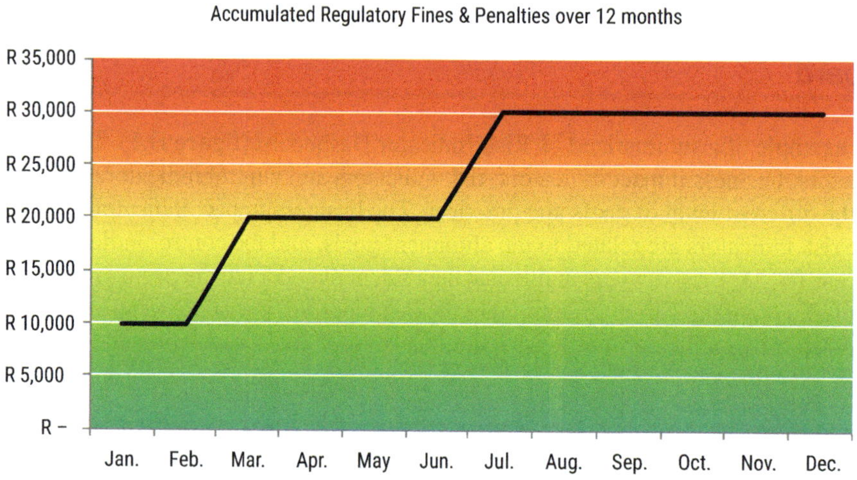

Figure D.3 is an example of a trending graph of IT risk incidents over 12 months. The organization's risk appetite and threshold inform the acceptable risk range (green), the early warning range (yellow/amber) and the unacceptable risk threshold (red).

Figure D.3—IT Risk Incident Metric

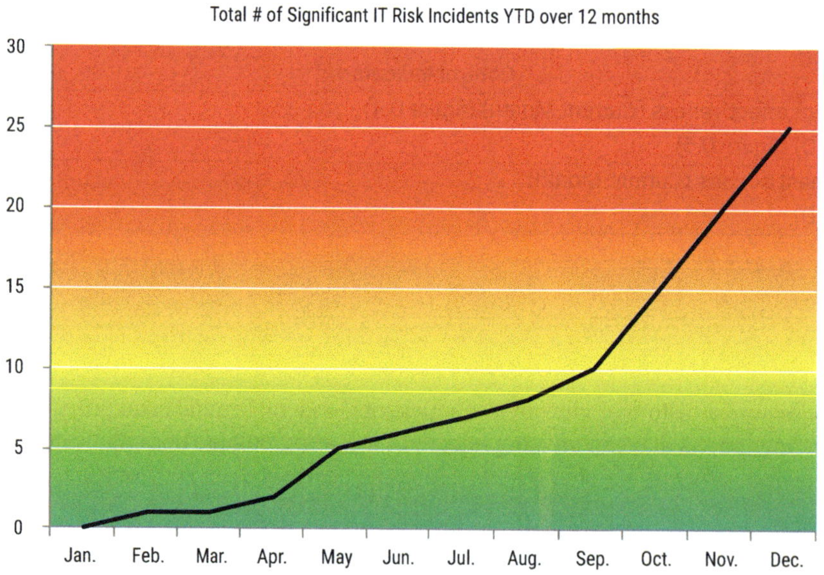

Figure D.4 is an example of a trending graph of the IT control environment effectiveness metric over 12 months. The enterprise's risk appetite and threshold inform the acceptable risk range (green), the early warning range (yellow/amber) and the unacceptable risk threshold (red).

Figure D.4—IT Control Effectiveness Metric

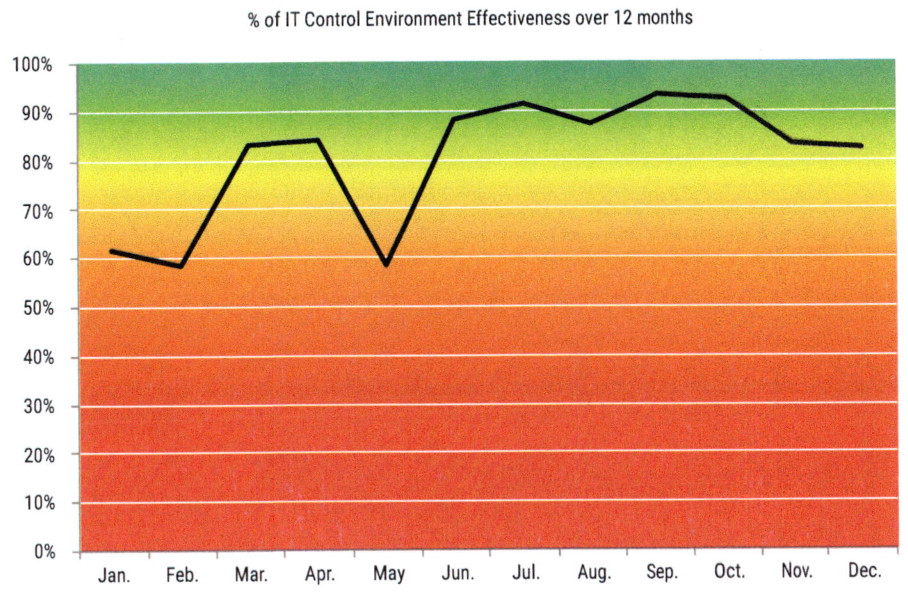

Figure D.5 is an example of a trending graph of the number of current IT control issues over 12 months. The enterprise's risk appetite and threshold inform the acceptable risk range (green), the early warning range (yellow/amber) and the unacceptable risk threshold (red).

Figure D.5—Current IT Control Issues Metric

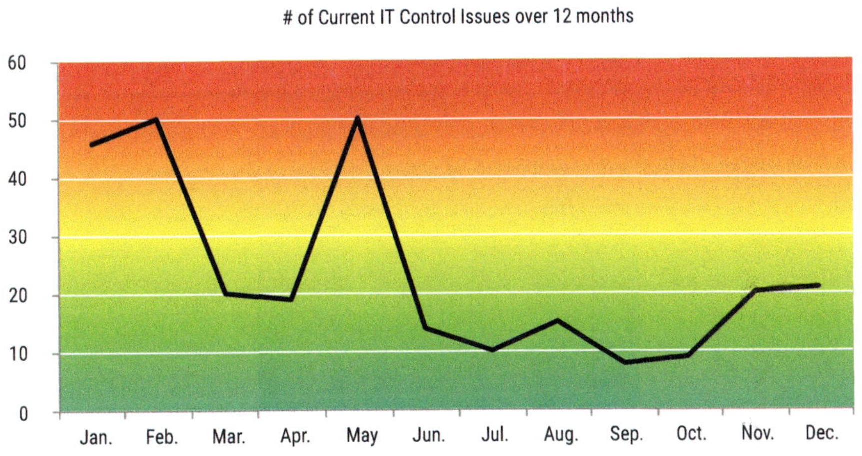

Page intentionally left blank

APPENDIX E
Sample Risk Maps

The following are example risk maps. In the examples, four levels of significance are defined:

- Red indicates unacceptable risk that requires immediate action
- Yellow/amber indicates unacceptable risk
- Green indicates an acceptable level of risk
- Blue indicates a very low risk

Figure E.1—Example Risk Map

	Frequency		
Impact	Unacceptable	Unacceptable—Immediate action required	Unacceptable—Immediate action required
	Acceptable	Unacceptable	Unacceptable—Immediate action required
	Opportunity	Acceptable	Unacceptable

Source: ISACA, *Risk IT Practitioner Guide, 2nd Edition*, USA, 2020, figure 3.3, *https://www.isaca.org/bookstore/bookstore-risk-digital/ritpg2*

Figure E.2—Risk Map Indicating Risk Appetite Bands

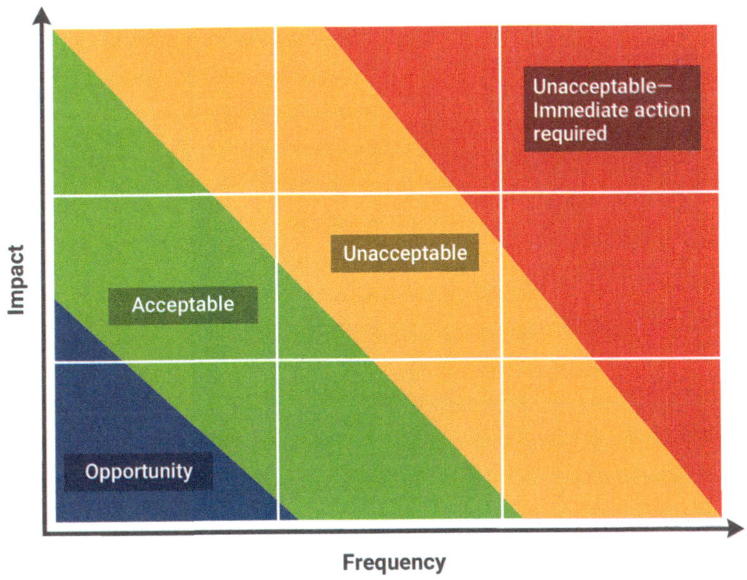

Source: ISACA, *Risk IT Practitioner Guide, 2nd Edition*, USA, 2020, figure 3.4, *https://www.isaca.org/bookstore/bookstore-risk-digital/ritpg2*

Page intentionally left blank

9781604208665